NANOENGINEERING: SCIENCE AND TECHNOLOGY

Edited by
Jurex Cuenca Gallo

Delve Publishing

Nanoengineering: Science and Technology

Edited by Jurex Cuenca Gallo

Edition 2018

Delve Publishing

Published by
Delve Publishing LLC
708 3rd Avenue, 6th Floor
New York
NY 10801
United States of America
www.delvepublishing.com

ISBN: 978-1-68095-910-9 (Hardback)

Printed in United States of America

Library of Congress Control Number: 2016920676

About the Editor

Jurex Cuenca Gallo

Dr. Jurex Gallo was conferred with his PhD degree in Chemical Engineering from De La Salle University – Manila in 2014. His expertise revolved in the fields of Material Science and Environmental Engineering focused on industrial and hazardous waste treatment and management. Presently, Dr. Gallo is affiliated as Senior Science Research Specialist of the Department of Science and Technology e-Asia Joint Research Program in the Philippines on developing functional carbon-based nanomaterials such as carbon nanotubes and graphene for energy and environmental applications.

Table of Contents

2 NANOPARTICLE

3 NANOELECTRONICS

8 MOLECULAR NANOTECHNOLOGY

9 NANOMECHANICS

List of Abbreviations

ATI	Above Threshold Ionization
AGIL	All Gas-Phase Iodine Laser
ASE	Amplified Spontaneous Emission
ASE	Amplified spontaneous emission
ADC	Analog To Digital Converter
APCI	Atmospheric-Pressure Chemical Ionization
AAS	Atomic absorption spectroscopy
CRLAS	Cavity Ring-Down Laser Absorption Spectroscopy
CRDS	Cavity ring-down spectroscopy
CCD	Charge Coupled Device
COIL	Chemical Oxygen Iodine Laser
CPA	Chirped pulse amplification
CW	Continuous wave,
DOS	Density of optical states
DFAs	Doped fibre amplifiers
DPSS	Doubled diode-pumped solid-state
EIT	Electromagnetically induced transparency
EBW	Electron Beam Welding
ESI	Electrospray Ionization
EDFA	Erbium Doped Fibre Amplifier
FDML	Fourier domain mode locking
FSR	Free Spectral Range
FEL	Free-electron lasers
FARS	Frequency-Agile Rapid Scanning
FROG	Frequency-Resolved Optical Gating
GC	Gas Chromatography

GFP	Green fluorescent protein
GDD	Group Delay Dispersion,
HHG	High harmonic generation
HMPGI	Hybrid Multiple-Prism Near-Grazing-Incidence
ICF	Inertial confinement fusion
KLM	Kerr-lens mode-locking
LBW	Laser beam welding
LIBS	Laser-induced breakdown spectroscopy
LIF	laser-induced fluorescence,
LLNL	Lawrence Livermore National Laboratory,
MS	Mass Spectrometry
MALDI	Matrix-assisted laser desorption/ionization
MEMS	Microelectromechanical Systems
MIIPS	Multiphoton intrapulse interference phase scan
MPI	Multi-Photon Ionization
MPL	Multiple-Prism Littrow
NIMS	Nanostructure Imaging Mass Spectrometry
ND	Neutral density
NOPA	Non-Collinear Parametric Amplifier
NMR	Nuclear magnetic resonance
OPA	Optical Parametric Amplifier
OPO	Optical Parametric Oscillator
OC	Output Coupler
PMT	Photomultiplier Tube
PW	Power Petawatt
PAM	Programmable array microscopes
PLD	Pulsed Laser Deposition
REMPI	Resonance-enhanced multiphoton ionization
SEM	Scanning Electron Microscope
SHG	Second Harmonic Generation
SOAs	Semiconductor Optical Amplifiers
SNR	Signal-To-Noise-Ratio
SSDL	Solid state dye lasers
SLM	Spatial light modulator

STED	Stimulated emission depletion microscopy
SALDI	Surface-assisted laser desorption/ionization
TCSPC	Time-correlated single photon counting
TRPIS	Time-resolved photo-ion spectroscopy
VCSELs	Vertical cavity surface-emitting lasers
XRF	X-ray fluorescence

Preface

Nanoengineering is the practice of engineering on the nanoscale. It derives its name from the nanometre, a unit of measurement equalling one billionth of a meter. Nanoengineering is largely a synonym for nanotechnology, but emphasizes the engineering rather than the pure science aspects of the field. The application of nanotechnology to biological systems has great potential for improving health care, food science, homeland security and energy production. Nanotechnology can be thought of as the design and synthesis of functional systems at the molecular level, where sizes are measured in nanometers - a billionth of a meter. Nanotechnology has had profound effect on material science, from newer computer processors based on nanofabrication that run faster, to stronger and safer construction materials. The field of biology is naturally ripe for nanotechnology implementation, as the building blocks of our genetic code and the proteins critical to all living systems are literally molecular machines. The effect of nanotechnology on the food sector is expected to be wide-ranging. Researchers and industry experts envision better-quality and safer foods with enhanced nutritional and health benefits through nanotechnology. It is critical for food nanotechnology applications to demonstrate superior benefits that outweigh risks to ensure their acceptance by the food industry and the public. Limited information on the health and environmental risks associated with nanosize food materials has hampered immediate adoption in food applications and has triggered public interest and concern regarding the associated health risks of nanomaterials for food use. With the right science-based information on associated risks and benefits, however, the future of nanotechnology in foods is expected to be bright. Nanotechnology provides a flexible and powerful platform to address infections and drug-resistant microbes. Infections that result from surgery represent a significant burden on limited health care resources. Many musculoskeletal surgeries, such as total joint replacement and trauma repair, involve the use of permanent or temporary implants that can provide vectors for infection. As the demand for such surgery increases from an aging population with traumatic injuries, the incidence of subsequent complications increases as well. Besides being small and allowing more transistors to be packed into a single chip, the uniform and symmetrical structure of nanotubes allows a higher electron mobility (faster electron movement in the material), a higher dielectric constant (faster frequency), and a symmetrical electron/hole characteristic. This text outlines a selection of exciting advances currently being made worldwide in the field of modern engineering at the nanometer scale.

Content Coverage

Chapters One and *Two* present an overview of nanomaterials and nanoparticles, correspondingly. Nanomaterials are slowly becoming commercialized and beginning to emerge as commodities. The study of nanoparticles is intense as they have many potential applications in medicine, optics, and electronics.

Chapters Three and *Four* present a coverage of nanoelectronics and bionanotechnology, consistently. Nanoelectronics covers a diverse set of devices and materials, with the common characteristic that they are so small that inter-atomic interactions and quantum mechanical properties need to be studied extensively. Bionanotechnology allows scientists to imagine and create systems that can be used for biological research.

Chapters Five and *Six* are intended to focus on nanolithography and nanofibers, separately. Applications of nanolithography include Multigate devices such as Field effect transistors (FET), Quantum dots, Nanowires, Gratings, Zone plates and Photomasks, nanoelectromechanical systems (NEMS), or semiconductor integrated circuits (nanocircuitry). However, nanofibers are an exciting new class of material used for several value added applications such as medical, filtration, barrier, wipes, personal care, composite, garments, insulation, and energy storage.

Chapters Seven and *Eight* clarify the fundamentals of nanaotoxicology and Molecular nanotechnology, separately. Nanotoxicology deals with the study and application of toxicity of nanomaterials, while application of molecular nanotechnology is smart materials.

Chapter Nine is nanomechanics which deals with the study and application of fundamental mechanical properties of physical systems at the nanoscale, such as elastic, thermal and kinetic material properties.

CHAPTER 1

Nanomaterials

Chapter Contents

Nanomaterials represent a new class of materials, where ever-smaller length scales (<100 nm) impart altered or enhanced properties. Two challenges in this area are (i) control of the structure of nanomaterials and (ii) controlling and defining the spatial placement of such materials. Nanomaterials research at Yale focuses on both these challenges through the synthesis and assembly of new hard and soft materials, approaching sizes as small as the molecular- to atomic scale. Whether created from self-assembly, chemical vapor deposition, solvothermal synthesis, or nanofabrication, these nanomaterials are poised to make great advances in the areas of solar cells, fuel cells, optical computing, biological science, catalysis, biomass conversion, and smart materials.

INTRODUCTION

Nanomaterials describe, in principle, materials of which a single unit is sized (in at least one dimension) between 1 and 1000 nanometres (10^{-9} meter) but is usually 1—100 nm (the usual definition of nanoscale).

Nanomaterials research takes a materials science-based approach to nanotechnology, leveraging advances in materials metrology and synthesis which have been developed in support of microfabrication research. Materials with structure at the nanoscale often have unique optical, electronic, or mechanical properties.

Nanomaterials are slowly becoming commercialized and beginning to emerge as commodities.

TYPES

Natural Nanomaterials

Biological systems often feature natural, functional nanomaterials. The structure of foraminifera (mainly chalk) and viruses (protein, capsid), the wax crystals covering a lotus or nasturtium leaf, spider and spider-mite silk, the blue hue of tarantulas, the "spatulae" on the bottom of gecko feet, some butterfly wing scales, natural colloids (milk, blood), horny materials (skin, claws, beaks, feathers, horns, hair), paper, cotton, nacre, corals, and even our own bone matrix are all natural *organic* nanomaterials.

Natural *inorganic* nanomaterials occur through crystal growth in the diverse chemical conditions of the Earth's crust. For example, clays display complex nanostructures due to anisotropy of their underlying crystal structure, and volcanic activity can give rise to opals, which are an instance of a naturally occurring photonic crystals due to their nanoscale structure. Fires represent particularly complex reactions and can produce pigments, cement, fumed silica etc.

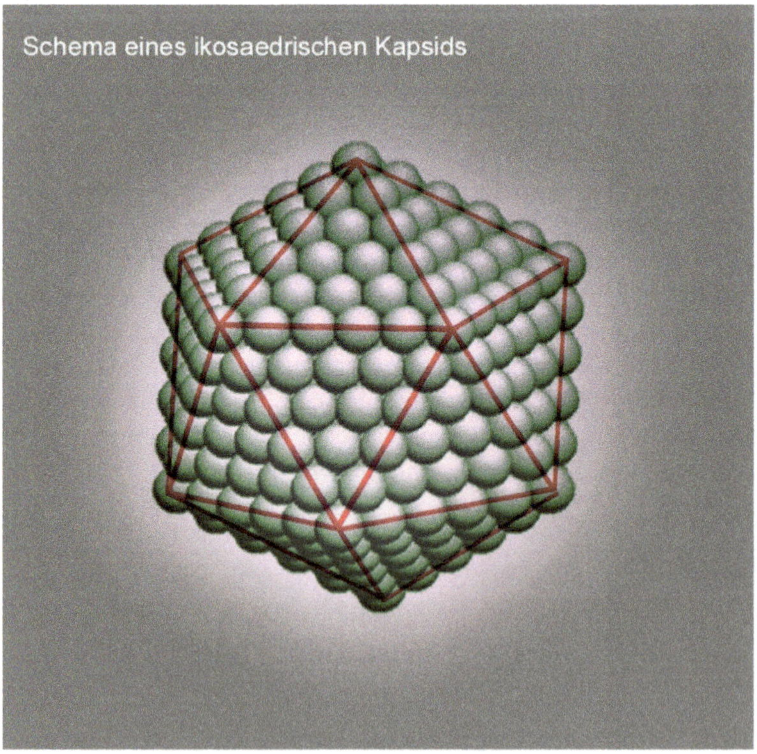

Viral capsid

"Lotus effect", hydrophobic effect with self-cleaning ability

Close-up of the underside of a gecko's foot as it walks on a glass wall. (spatula: 200 × 10-15 nm).

SEM micrograph of a butterfly wing scale (× 5000)

Peacock feather (detail)

Brazilian Crystal Opal. The play of color is caused by the interference and diffraction of light between silica spheres (150 - 300 nm in diameter).

Lycurgus Cup, glass, 4th century, Roman. Nanoparticles (70 nm) of gold and silver, dispersed in colloidal form, are responsible for the dichroic effect (red/green).

Blue hue of a species of tarantula (450 nm ± 20 nm).

Fullerenes

The fullerenes are a class of allotropes of carbon which conceptually are graphene sheets rolled into tubes or spheres. These include the carbon nanotubes (or silicon nanotubes) which are of interest both because of their mechanical strength and also because of their electrical properties.

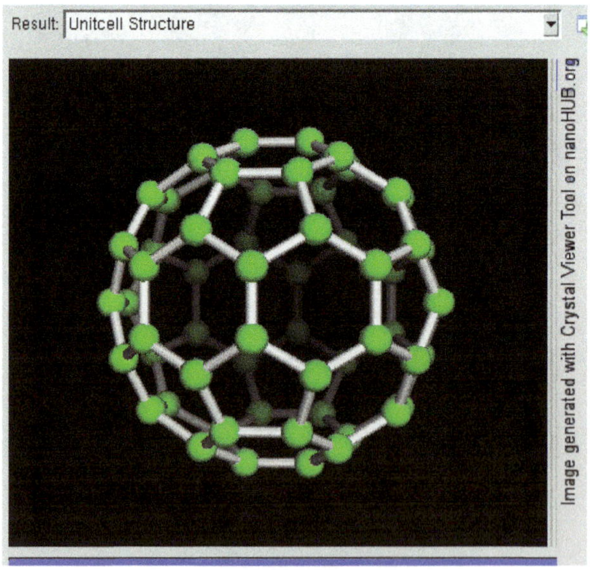

Rotating view of C_{60}, one kind of fullerene.

The first fullerene molecule to be discovered, and the family's namesake, buckminsterfullerene (C_{60}), was prepared in 1985 by Richard Smalley, Robert Curl, James Heath, Sean O'Brien, and

Harold Kroto at Rice University. The name was a homage to Buckminster Fuller, whose geodesic domes it resembles. Fullerenes have since been found to occur in nature. More recently, fullerenes have been detected in outer space.

For the past decade, the chemical and physical properties of fullerenes have been a hot topic in the field of research and development, and are likely to continue to be for a long time. In April 2003, fullerenes were under study for potential medicinal use: binding specific antibiotics to the structure of resistant bacteria and even target certain types of cancer cells such as melanoma. The October 2005 issue of Chemistry and Biology contains an article describing the use of fullerenes as light-activated antimicrobial agents. In the field of nanotechnology, heat resistance and superconductivity are among the properties attracting intense research.

A common method used to produce fullerenes is to send a large current between two nearby graphite electrodes in an inert atmosphere. The resulting carbon plasma arc between the electrodes cools into sooty residue from which many fullerenes can be isolated.

There are many calculations that have been done using ab-initio Quantum Methods applied to fullerenes. By DFT and TDDFT methods one can obtain IR, Raman and UV spectra. Results of such calculations can be compared with experimental results.

Graphene Nanostructures

2D materials are crystalline materials consisting of a two-dimensional single layer of atoms. The most important representative graphene was discovered in 2004. Other 2D materials based on other elements have since been reported. Box-shaped graphene (BSG) nanostructure is an example of 3D nanomaterial. BSG nanostructure has appeared after mechanical cleavage of pyrolytic graphite. This nanostructure is a multilayer system of parallel hollow nanochannels located along the surface and having quadrangular cross-section. The thickness of the channel walls is approximately equal to 1 nm. The typical width of channel facets makes about 25 nm.

NANOPARTICLES

Inorganic nanomaterials, (e.g. quantum dots, nanowires and nanorods) because of their interesting optical and electrical properties, could be used in optoelectronics. Furthermore, the optical and electronic properties of nanomaterials which depend on their size and shape can be tuned via synthetic techniques. There are the possibilities to use those materials in organic material based optoelectronic devices such as Organic solar cells, OLEDs etc. The operating principles of such devices are governed by photoinduced processes like electron transfer and energy transfer. The performance of the devices depends on the efficiency of the photoinduced process responsible for their functioning. Therefore, better understanding of those photoinduced processes in organic/inorganic nanomaterial composite systems is necessary in order to use them in organic optoelectronic devices. Nanoparticles or nanocrystals made of metals, semiconductors, or oxides are of particular interest for their mechanical, electrical, magnetic, optical, chemical and other properties. Nanoparticles have been used as quantum dots and as chemical catalysts such as nanomaterial-based catalysts. Recently, a range of nanoparticles are extensively investigated for biomedical applications including tissue engineering, drug delivery, biosensor.

Nanoparticles are of great scientific interest as they are effectively a bridge between bulk materials and atomic or molecular structures. A bulk material should have constant physical properties regardless of its size, but at the nano-scale this is often not the case. Size-dependent properties are observed such as quantum confinement in semiconductor particles, surface plasmon resonance in some metal particles and superparamagnetism in magnetic materials. Nanoparticles exhibit a number of special properties relative to bulk material. For example, the bending of bulk copper (wire, ribbon, etc.) occurs with movement of copper atoms/clusters at about the 50 nm scale. Copper nanoparticles smaller than 50 nm are considered super hard materials that do not exhibit the same malleability and ductility as bulk copper. The change in properties is not always desirable. Ferroelectric materials smaller than 10 nm can switch their magnetisation direction using room temperature thermal energy, thus making them useless for memory storage. Suspensions of nanoparticles are possible because the interaction of the particle surface with the solvent is strong enough to overcome differences in density, which usually result in a material either sinking or floating in a liquid. Nanoparticles often have unexpected visual properties because they are small enough to confine their electrons and produce quantum effects. For example, gold nanoparticles appear deep red to black in solution.

The often very high surface area to volume ratio of nanoparticles provides a tremendous driving force for diffusion, especially at elevated temperatures. Sintering is possible at lower temperatures and over shorter durations than for larger particles. This theoretically does not affect the density of the final product, though flow difficulties and the tendency of nanoparticles to agglomerate do complicate matters. The surface effects of nanoparticles also reduces the incipient melting temperature.

NANOZYMES

Nanozymes are nanomaterials with enzyme-like characteristics. They are an emerging type of artificial enzyme, which have been used for wide applications in such as biosensing, bioimaging, tumor diagnosis and therapy, antibiofouling, etc.

SYNTHESIS

The goal of any synthetic method for nanomaterials is to yield a material that exhibits properties that are a result of their characteristic length scale being in the nanometer range (~1 – 100 nm). Accordingly, the synthetic method should exhibit control of size in this range so that one property or another can be attained. Often the methods are divided into two main types "Bottom Up" and "Top Down."

Bottom up Methods

Bottom up methods involve the assembly of atoms or molecules into nanostructured arrays. In these methods the raw material sources can be in the form of gases, liquids or solids. The latter requiring some sort of disassembly prior to their incorporation onto a nanostructure. Bottom methods generally fall into two categories: chaotic and controlled.

Chaotic Processes

Chaotic processes involve elevating the constituent atoms or molecules to a chaotic state and then suddenly changing the conditions so as to make that state unstable. Through the clever manipulation of any number of parameters, products form largely as a result of the insuring kinetics. The collapse from the chaotic state can be difficult or impossible to control and so ensemble statistics often govern the resulting size distribution and average size. Accordingly, nanoparticle formation is controlled through manipulation of the end state of the products.

Examples of Chaotic Processes are: Laser ablation, Exploding wire, Arc, Flame pyrolysis, Combustion, Precipitation synthesis techniques.

Controlled Processes

Controlled Processes involve the controlled delivery of the constituent atoms or molecules to the site(s) of nanoparticle formation such that the nanoparticle can grow to a prescribed sizes in a controlled manner. Generally the state of the constituent atoms or molecules are never far from that needed for nanoparticle formation. Accordingly, nanoparticle formation is controlled through the control of the state of the reactants.

Examples of controlled processes are self-limiting growth solution, self-limited chemical vapor deposition, shaped pulse femtosecond laser techniques, and molecular beam epitaxy.

CHARACTERIZATION

Novel effects can occur in materials when structures are formed with sizes comparable to any one of many possible length scales, such as the de Broglie wavelength of electrons, or the optical wavelengths of high energy photons. In these cases quantum mechanical effects can dominate material properties. One example is quantum confinement where the electronic properties of solids are altered with great reductions in particle size. The optical properties of nanoparticles, e.g. fluorescence, also become a function of the particle diameter. This effect does not come into play by going from macrosocopic to micrometer dimensions, but becomes pronounced when the nanometer scale is reached.

In addition to optical and electronic properties, the novel mechanical properties of many nanomaterials is the subject of nanomechanics research. When added to a bulk material, nanoparticles can strongly influence the mechanical properties of the material, such as the stiffness or elasticity. For example, traditional polymers can be reinforced by nanoparticles (such as carbon nanotubes) resulting in novel materials which can be used as lightweight replacements for metals. Such composite materials may enable a weight reduction accompanied by an increase in stability and improved functionality.

Finally, nanostructured materials with small particle size such as zeolites, and asbestos, are used as catalysts in a wide range of critical industrial chemical reactions. The further development of such catalysts can form the basis of more efficient, environmentally friendly chemical processes.

The first observations and size measurements of nano-particles were made during the first decade of the 20th century. Zsigmondy made detailed studies of gold sols and other nanomaterials with sizes down to 10 nm and less. He published a book in 1914. He used an ultramicroscope that employs a *dark field* method for seeing particles with sizes much less than light wavelength. There are traditional techniques developed during 20th century in Interface and Colloid Science for characterizing nanomaterials. These are widely used for *first generation* passive nanomaterials specified in the next section.

These methods include several different techniques for characterizing particle size distribution. This characterization is imperative because many materials that are expected to be nano-sized are actually aggregated in solutions. Some of methods are based on light scattering. Others apply ultrasound, such as ultrasound attenuation spectroscopy for testing concentrated nano-dispersions and microemulsions.

There is also a group of traditional techniques for characterizing surface charge or zeta potential of nano-particles in solutions. This information is required for proper system stabilzation, preventing its aggregation or flocculation. These methods include microelectrophoresis, electrophoretic light scattering and electroacoustics. The last one, for instance colloid vibration current method is suitable for characterizing concentrated systems.

UNIFORMITY

The chemical processing and synthesis of high performance technological components for the private, industrial and military sectors requires the use of high purity ceramics, polymers, glass-ceramics and material composites. In condensed bodies formed from fine powders, the irregular sizes and shapes of nanoparticles in a typical powder often lead to non-uniform packing morphologies that result in packing density variations in the powder compact. Uncontrolled agglomeration of powders due to attractive van der Waals forces can also give rise to in microstructural inhomogeneities. Differential stresses that develop as a result of non-uniform drying shrinkage are directly related to the rate at which the solvent can be removed, and thus highly dependent upon the distribution of porosity. Such stresses have been associated with a plastic-to-brittle transition in consolidated bodies, and can yield to crack propagation in the unfired body if not relieved.

In addition, any fluctuations in packing density in the compact as it is prepared for the kiln are often amplified during the sintering process, yielding inhomogeneous densification. Some pores and other structural defects associated with density variations have been shown to play a detrimental role in the sintering process by growing and thus limiting end-point densities. Differential stresses arising from inhomogeneous densification have also been shown to result in the propagation of internal cracks, thus becoming the strength-controlling flaws.

It would therefore appear desirable to process a material in such a way that it is physically uniform with regard to the distribution of components and porosity, rather than using particle size distributions which will maximize the green density. The containment of a uniformly dispersed assembly of strongly interacting particles in suspension requires total control over particle-particle interactions. It should be noted here that a number of dispersants such as ammonium citrate (aqueous) and imidazoline or oleyl alcohol (nonaqueous) are promising solutions as possible additives for enhanced dispersion and deagglomeration. Monodisperse nanoparticles and colloids provide this potential.

Monodisperse powders of colloidal silica, for example, may therefore be stabilized sufficiently to ensure a high degree of order in the colloidal crystal or polycrystalline colloidal solid which results from aggregation. The degree of order appears to be limited by the time and space allowed for longer-range correlations to be established. Such defective polycrystalline colloidal structures would appear to be the basic elements of sub-micrometer colloidal materials science, and, therefore, provide the first step in developing a more rigorous understanding of the mechanisms involved in microstructural evolution in high performance materials and components.

LEGAL DEFINITION

On 18 October 2011, the European Commission adopted the following definition of a nanomaterial: "A natural, incidental or manufactured material containing particles, in an unbound state or as an aggregate or as an agglomerate and where, for 50% or more of the particles in the number size distribution, one or more external dimensions is in the size range 1 nm – 100 nm. In specific cases and where warranted by concerns for the environment, health, safety or competitiveness the number size distribution threshold of 50% may be replaced by a threshold between 1 and 50%." However, this differs from the definition adopted by the International Organization for Standardization (ISO), which is: "Material with any external dimension in the nanoscale or having internal structure in the nanoscale."

"Nanoscale" is, in turn, defined as: "Size range from approximately 1 nm to 100 nm."

It is not currently known which of these, if any, will prevail in courts of law.

SAFETY

Health impact

While nanomaterials and nanotechnologies are expected to yield numerous health and health care advances, such as more targeted methods of delivering drugs, new cancer therapies, and methods of early detection of diseases, they also may have unwanted effects. Increased rate of absorption is the main concern associated with manufactured nanoparticles.

When materials are made into nanoparticles, their surface area to volume ratio increases. The greater specific surface area (surface area per unit weight) may lead to increased rate of absorption through the skin, lungs, or digestive tract and may cause unwanted effects to the lungs as well as other organs. However, the particles must be absorbed in sufficient quantities in order to pose health risks.

Nanoparticles created adventitiously (e.g., through the rubbing of prostheses) have long been known to be a health hazard, but as the use of nanomaterials increases worldwide, concerns for worker and user safety are mounting. To address such concerns, the Swedish Karolinska Institute conducted a study in which various nanoparticles were introduced to human lung epithelial cells. The results, released in 2008, showed that iron oxide nanoparticles caused little DNA damage and were non-toxic. Zinc oxide nanoparticles were slightly worse. Titanium dioxide caused only DNA damage. Carbon nanotubes caused DNA damage at low levels. Copper oxide was found to be the worst offender, and was the only nanomaterial identified by the researchers as a clear health risk. Though nanomaterials are not confirmed as a health risk to workers who produce them, NIOSH recommends that exposure precautions and personal protective equipment be used to protect workers until the risks of nanomaterial manufacture are better understood.

Occupational Safety

Beyond normal chemical safety practices such as using personal protective equipment, The United States National Institute for Occupational Safety and Health recommends the use of HEPA filtration on local exhaust ventilation such as fume hoods. General laboratory ventilation is often not sufficient to effectively clear nanomaterials released into the general room air over a 30-minute period; therefore, researchers should leave the hood fan on even after synthesis is complete. Nanomaterials in dry powder form have the most risk of airborne dust generation and dermal contact, followed by liquid suspensions and then by nanomaterials embedded in a solid matrix. Nanoparticles behave differently than other similarly sized particles. It is therefore necessary to develop specialized approaches to testing and monitoring their effects on human health and on the environment. The OECD Chemicals Committee has established the Working Party on Manufactured Nanomaterials to address this issue and to study the practices of OECD member countries in regards to nanomaterial safety.

CARBON NANOTUBES

Carbon nanotubes (CNTs) are allotropes of carbon with a cylindricalnanostructure. These cylindrical carbon molecules have unusual properties, which are valuable for nanotechnology, electronics, optics and other fields of materials science and technology. Owing to the material's exceptional strength and stiffness, nanotubes have been constructed with length-to-diameter ratio of up to 132,000,000:1, significantly larger than for any other material. In addition, owing to their extraordinary thermal conductivity, mechanical, and electrical properties, carbon nanotubes find applications as additives to various structural materials. For instance, nanotubes form a tiny portion of the material(s) in some (primarily carbon fiber) baseball bats, golf clubs, car parts or damascus steel. Nanotubes are members of the fullerene structural family. Their name is derived from their long, hollow structure with the walls formed by one-atom-thick sheets of carbon, called graphene. These sheets are rolled at specific and discrete ("chiral") angles, and the combination of the rolling angle and radius decides the nanotube properties; for example, whether the individual nanotube shell is a metal or semiconductor. Nanotubes are categorized as single-walled nanotubes (SWNTs) and multi-walled nanotubes (MWNTs). Individual nanotubes naturally align themselves into "ropes" held together by van der Waals forces, more specifically, pi-stacking. Applied quantum chemistry, specifically, orbital hybridization best describes chemical bonding in nanotubes. The chemical bonding of nanotubes is composed entirely of sp^2 bonds, similar to those of graphite. These bonds, which are stronger than the sp^3 bonds found in alkanes and diamond, provide nanotubes with their unique strength.

Types of Carbon Nanotubes and Related Structures

Terminology

There is no consensus on some terms describing carbon nanotubes in scientific literature: both "-wall" and "-walled" are being used in combination with "single", "double", "triple" or "multi", and the letter C is often omitted in the abbreviation; for example, multi-walled carbon nanotube (MWNT).

Single-walled

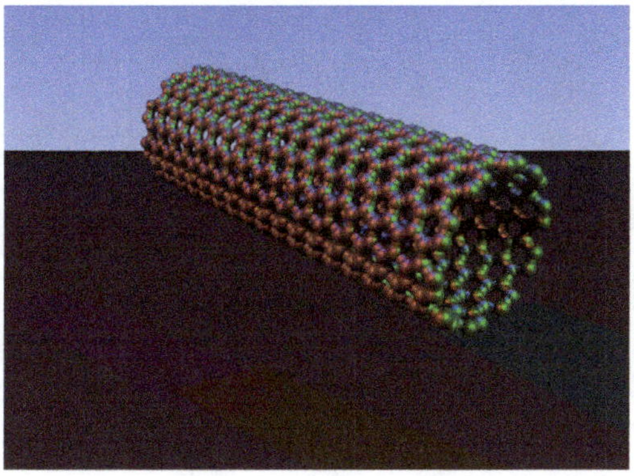

Armchair (*n*,*n*) i.e.: m=n.

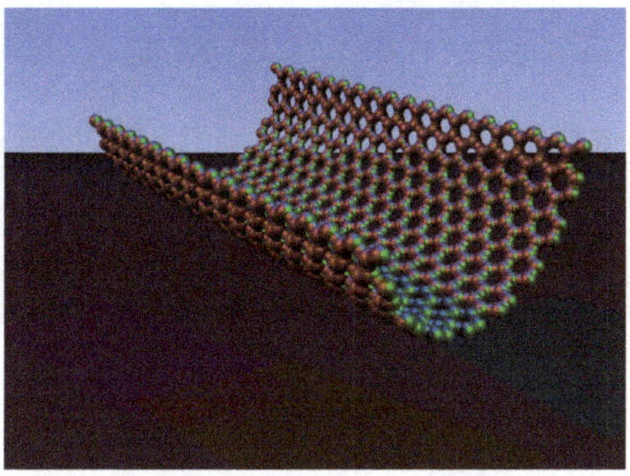

The translation vector is bent, while the chiral vector stays straight.

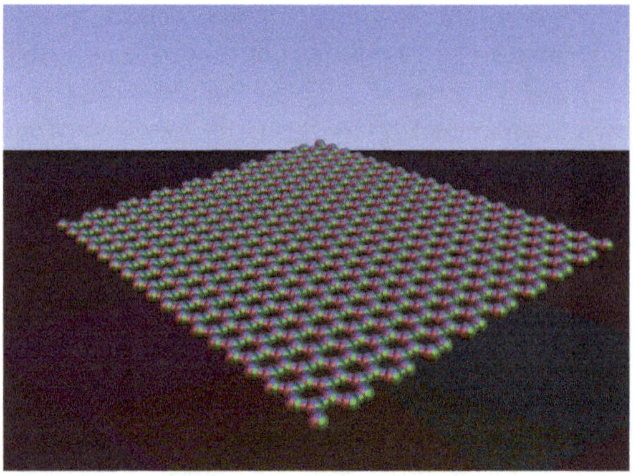

Graphene nanoribbon.

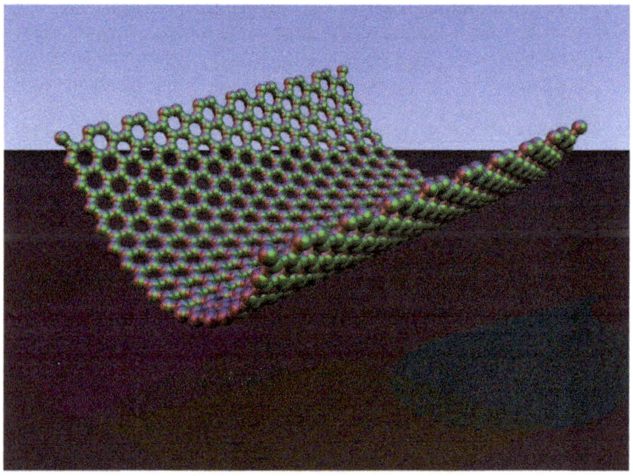

The chiral vector is bent, while the translation vector stays straight.

Zigzag (n,0).

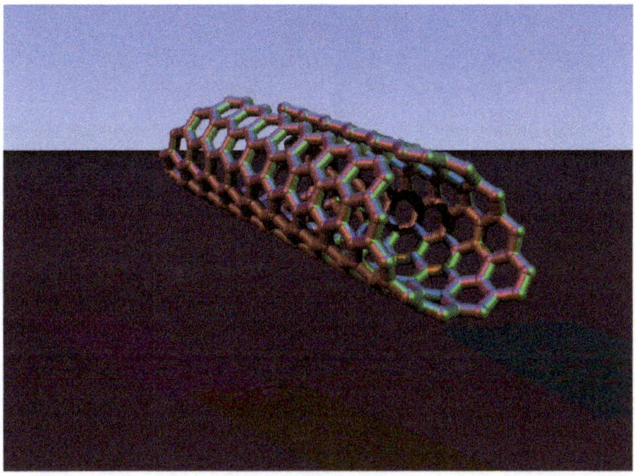

Chiral (n,m).

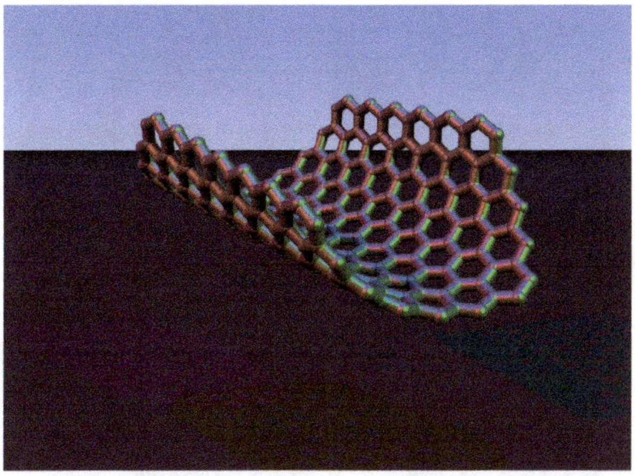

n and m can be counted at the end of the tube..

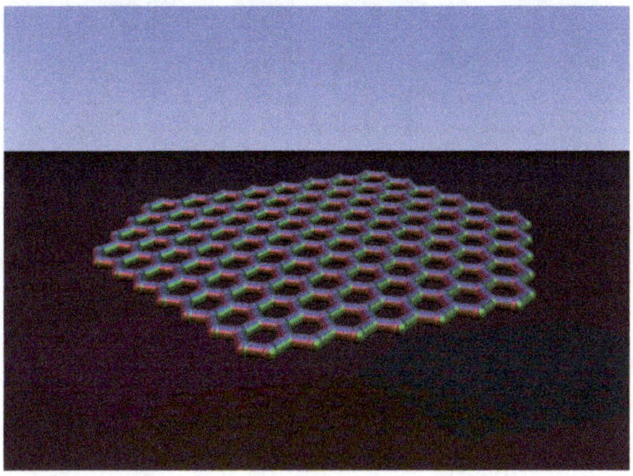

Graphene nanoribbon.

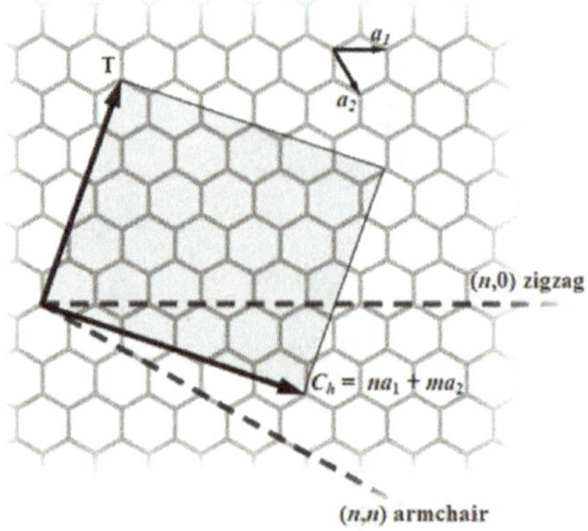

The (n,m) nanotube naming scheme can be thought of as a vector ($\mathbf{C}_h$) in an infinite graphene sheet that describes how to "roll up" the graphene sheet to make the nanotube. **T** denotes the tube axis, and $\mathbf{a}_1$ and $\mathbf{a}_2$ are the unit vectors of graphene in real space.

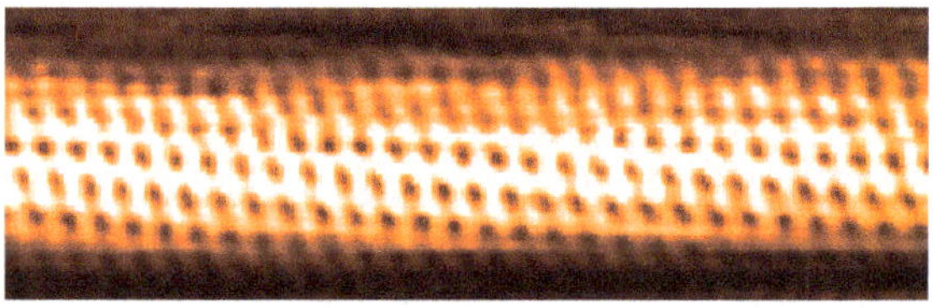

A scanning tunneling microscopy image of single-walled carbon nanotube.

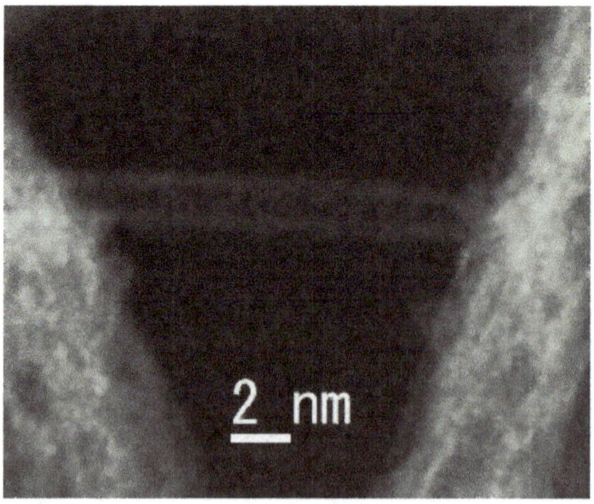

A transmission electron microscopy image of a single-walled carbon nanotube.

Most single-walled nanotubes (SWNTs) have a diameter of close to 1 nanometer, and can be many millions of times longer. The structure of a SWNT can be conceptualized by wrapping a one-atom-thick layer of graphite called graphene into a seamless cylinder.

The way the graphene sheet is wrapped is represented by a pair of indices (*n*,*m*). The integers *n* and *m* denote the number of unit vectors along two directions in the honeycomb crystal lattice of graphene. If $m = 0$, the nanotubes are called **zigzag** nanotubes, and if $n = m$, the nanotubes are called **armchair** nanotubes. Otherwise, they are called **chiral**. The diameter of an ideal nanotube can be calculated from its (n,m) indices as follows

$$d = \frac{a}{\pi}\sqrt{(n^2 + nm + m^2)} = 78.3\sqrt{((n+m)^2 - nm)}\text{pm},$$

where $a = 0.246$ nm.

SWNTs are an important variety of carbon nanotube because most of their properties change significantly with the (*n*,*m*) values, and this dependence is non-monotonic (see Kataura plot). In particular, their band gap can vary from zero to about 2 eV and their electrical conductivity can show metallic or semiconducting behavior. Single-walled nanotubes are likely candidates for miniaturizing electronics. The most basic building block of these systems is the electric wire, and SWNTs with diameters of an order of a nanometer can be excellent conductors. One useful application of SWNTs is in the development of the first intermolecular field-effect transistors (FET). The first intermolecular logic gate using SWCNT FETs was made in 2001. A logic gate requires both a p-FET and an n-FET. Because SWNTs are p-FETs when exposed to oxygen and n-FETs otherwise, it is possible to expose half of an SWNT to oxygen and protect the other half from it. The resulting SWNT acts as a *not* logic gate with both p and n-type FETs in the same molecule.

Prices for single-walled nanotubes declined from around $1500 per gram as of 2000 to retail prices of around $50 per gram of as-produced 40–60% by weight SWNTs as of March 2010. As of 2016 the retail price of as-produced 75% by weight SWNTs were $2 per gram, cheap enough for widespread use. SWNTs are forecast to make a large impact in electronics applications by 2020 according to the *The Global Market for Carbon Nanotubes* report.

Multi-walled

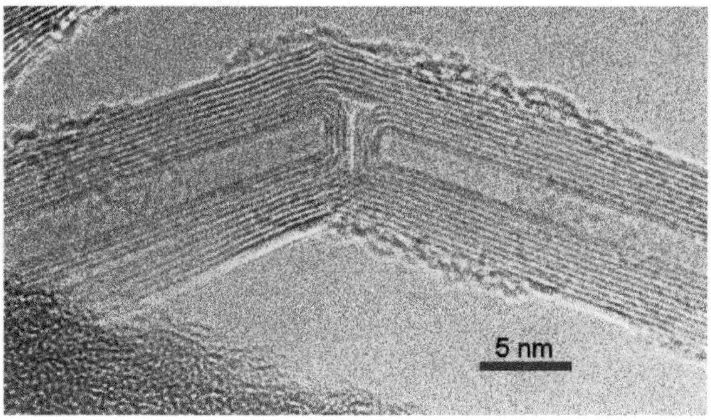

A scanning electron microscopy image of carbon nanotubes bundles.

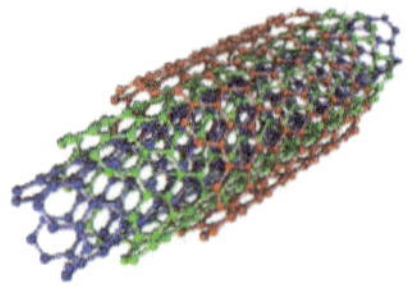

Triple-walled armchair carbon nanotube.

Multi-walled nanotubes (MWNTs) consist of multiple rolled layers (concentric tubes) of graphene. There are two models that can be used to describe the structures of multi-walled nanotubes. In the *Russian Doll* model, sheets of graphite are arranged in concentric cylinders, e.g., a (0,8) single-walled nanotube (SWNT) within a larger (0,17) single-walled nanotube. In the *Parchment* model, a single sheet of graphite is rolled in around itself, resembling a scroll of parchment or a rolled newspaper. The interlayer distance in multi-walled nanotubes is close to the distance between graphene layers in graphite, approximately 3.4 Å. The Russian Doll structure is observed more commonly. Its individual shells can be described as SWNTs, which can be metallic or semiconducting. Because of statistical probability and restrictions on the relative diameters of the individual tubes, one of the shells, and thus the whole MWNT, is usually a zero-gap metal Double-walled carbon nanotubes (DWNTs) form a special class of nanotubes because their morphology and properties are similar to those of SWNTs but they are more resistant to chemicals. This is especially important when it is necessary to graft chemical functions to the surface of the nanotubes (functionalization) to add properties to the CNT. Covalent functionalization of SWNTs will break some C=C double bonds, leaving "holes" in the structure on the nanotube, and thus modifying both its mechanical and electrical properties. In the case of DWNTs, only the outer wall is modified. DWNT synthesis on the gram-scale was first proposed in 2003 by the CCVD technique, from the selective reduction of oxide solutions in methane and hydrogen. The telescopic motion ability of inner shells and their unique mechanical properties will permit the use of multi-walled nanotubes as main movable arms in coming nanomechanical devices. Retraction force that occurs to telescopic motion caused by the Lennard-Jones interaction between shells and its value is about 1.5 nN.

Nanotube Junctions

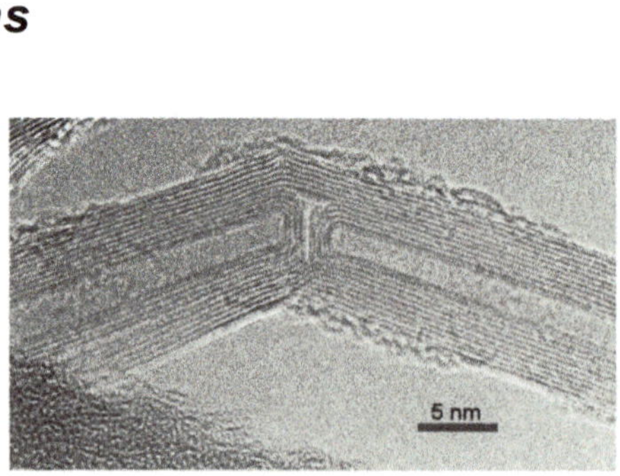

Transmission electron microscope image of carbon nanotube junction.

Junctions between 2 or more nanotubes have been widely discussed theoretically. Such junctions are quite frequently observed in samples prepared by arc discharge as well as by Chemical vapor deposition. The electronic properties of such junctions were first considered theoretically by Lambin et al., who pointed out that a connection between metallic tube and a semiconducting one would represent a nanoscale heterojunction. Such a junction could therefore form a component of a nanotube-based electronic circuit. The image on the right shows a junction between two multiwalled nanotubes.

Junctions between nanotubes and graphene have been considered theoretically,but not widely studied experimentally. Such junctions form the basis of pillared graphene, in which parallel graphene sheets are separated by short nanotubes. Pillared graphene represents a class of three-dimensional carbon nanotube architectures.

Torus

In theory, a nanotorus is a carbon nanotube bent into a torus (doughnut shape). Nanotori are predicted to have many unique properties, such as magnetic moments 1000 times larger than previously expected for certain specific radii. Properties such as magnetic moment, thermal stability, etc. vary widely depending on radius of the torus and radius of the tube.

Nanobud

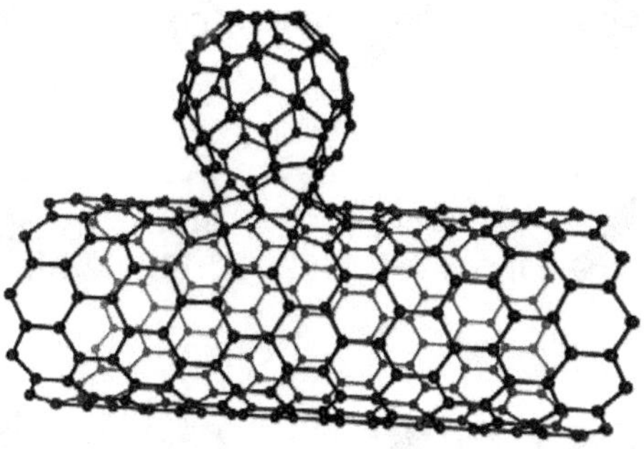

A stable nanobud structure.

Carbon nanobuds are a newly created material combining two previously discovered allotropes of carbon: carbon nanotubes and fullerenes. In this new material, fullerene-like "buds" are covalently bonded to the outer sidewalls of the underlying carbon nanotube. This hybrid material has useful properties of both fullerenes and carbon nanotubes. In particular, they have been found to be exceptionally good field emitters. In composite materials, the attached fullerene molecules may function as molecular anchors preventing slipping of the nanotubes, thus improving the composite's mechanical properties.

Three-dimensional Carbon Nanotube Architectures

3D carbon scaffolds.

Recently, several studies have highlighted the prospect of using carbon nanotubes as building blocks to fabricate three-dimensional macroscopic (>100 nm in all three dimensions) all-carbon devices. Lalwani et al. have reported a novel radical initiated thermal crosslinking method to fabricate macroscopic, free-standing, porous, all-carbon scaffolds using single- and multi-walled carbon nanotubes as building blocks. These scaffolds possess macro-, micro-, and nano- structured pores and the porosity can be tailored for specific applications. These 3D all-carbon scaffolds/architectures may be used for the fabrication of the next generation of energy storage, supercapacitors, field emission transistors, high-performance catalysis, photovoltaics, and biomedical devices and implants.

Graphenated Carbon Nanotubes (g-CNTs)

SEM series of graphenated CNTs with varying foliate density

Graphenated CNTs are a relatively new hybrid that combines graphitic foliates grown along the sidewalls of multiwalled or bamboo style CNTs. Yu *et al.* reported on "chemically bonded graphene leaves" growing along the sidewalls of CNTs. Stoner *et al.* described these structures as "graphenated CNTs" and reported in their use for enhanced supercapacitor performance. Hsu *et al.* further reported on similar structures formed on carbon fiber paper, also for use in supercapacitor applications. Pham *et al.* also reported a similar structure, namely "graphene-carbon nanotube hybrids", grown directly onto carbon fiber paper to form an integrated, binder free, high surface area conductive catalyst support for Proton Exchange Membrane Fuel Cells electrode applications with enhanced performance and durability. The foliate density can vary as a function of deposition conditions (e.g. temperature and time) with their structure ranging from few layers of graphene (< 10) to thicker, more graphite-like.

The fundamental advantage of an integrated graphene-CNT structure is the high surface area three-dimensional framework of the CNTs coupled with the high edge density of graphene. Graphene edges provide significantly higher charge density and reactivity than the basal plane, but they are difficult to arrange in a three-dimensional, high volume-density geometry. CNTs are readily aligned in a high density geometry (i.e., a vertically aligned forest) but lack high charge density surfaces—the sidewalls of the CNTs are similar to the basal plane of graphene and exhibit low charge density except where edge defects exist. Depositing a high density of graphene foliates along the length of aligned CNTs can significantly increase the total charge capacity per unit of nominal area as compared to other carbon nanostructures.

Nitrogen-doped Carbon Nanotubes

Nitrogen doped carbon nanotubes (N-CNTs) can be produced through five main methods, chemical vapor deposition,high-temperature and high-pressure reactions, gas-solid reaction of amorphous carbon with NH_3 at high temperature, solid reaction, and solvothermal synthesis.

N-CNTs can also be prepared by a CVD method of pyrolyzing melamine under Ar at elevated temperatures of 800–980 °C. However synthesis by CVD of melamine results in the formation of bamboo-structured CNTs. XPS spectra of grown N-CNTs reveal nitrogen in five main components, pyridinic nitrogen, pyrrolic nitrogen, quaternary nitrogen, and nitrogen oxides. Furthermore, synthesis temperature affects the type of nitrogen configuration.

Nitrogen doping plays a pivotal role in lithium storage, as it creates defects in the CNT walls allowing for Li ions to diffuse into interwall space. It also increases capacity by providing more favorable bind of N-doped sites. N-CNTs are also much more reactive to metal oxide nanoparticle deposition which can further enhance storage capacity, especially in anode materials for Li-ion batteries. However boron-doped nanotubes have been shown to make batteries with triple capacity.

Peapod

A carbon peapod is a novel hybrid carbon material which traps fullerene inside a carbon nanotube. It can possess interesting magnetic properties with heating and irradiation. It can also be applied as an oscillator during theoretical investigations and predictions.

Cup-stacked Carbon Nanotubes

Cup-stacked carbon nanotubes (CSCNTs) differ from other quasi-1D carbon structures, which normally behave as quasi-metallic conductors of electrons. CSCNTs exhibit semiconducting behaviors due to the stacking microstructure of graphene layers.

Extreme Carbon Nanotubes

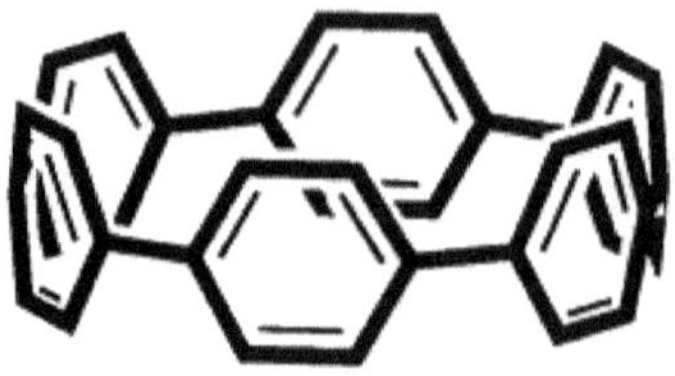

Cycloparaphenylene.

The observation of the *longest* carbon nanotubes grown so far are over 1/2 m (550 mm long) was reported in 2013. These nanotubes were grown on Si substrates using an improved chemical vapor deposition (CVD) method and represent electrically uniform arrays of single-walled carbon nanotubes.

The *shortest* carbon nanotube is the organic compound cycloparaphenylene, which was synthesized in 2008.

The *thinnest* carbon nanotube is the armchair (2,2) CNT with a diameter of 0.3 nm. This nanotube was grown inside a multi-walled carbon nanotube. Assigning of carbon nanotube type was done by a combination of high-resolution transmission electron microscopy (HRTEM), Raman spectroscopy and density functional theory (DFT) calculations.

The *thinnest freestanding* single-walled carbon nanotube is about 0.43 nm in diameter. Researchers suggested that it can be either (5,1) or (4,2) SWCNT, but the exact type of carbon nanotube remains questionable. (3,3), (4,3) and (5,1) carbon nanotubes (all about 0.4 nm in diameter) were unambiguously identified using aberration-corrected high-resolution transmission electron microscopy inside double-walled CNTs.

The *highest density* of CNTs was achieved in 2013, grown on a conductive titanium-coated copper surface that was coated with co-catalysts cobalt and molybdenum at lower than typical temperatures of 450 °C. The tubes averaged a height of 380 nm and a mass density of 1.6 g cm^{-3}. The material showed ohmic conductivity (lowest resistance ~22 kΩ).

Properties

Strength

Carbon nanotubes are the strongest and stiffest materials yet discovered in terms of tensile strength and elastic modulus respectively. This strength results from the covalent sp^2 bonds formed between the individual carbon atoms. In 2000, a multi-walled carbon nanotube was tested to have a tensile strength of 63 gigapascals (9,100,000 psi). (For illustration, this translates into the ability to endure tension of a weight equivalent to 6,422 kilograms-force (62,980 N; 14,160 lbf) on a cable with cross-section of 1 square millimetre (0.0016 sq in).) Further studies, such as one conducted in 2008, revealed that individual CNT shells have strengths of up to ≈100 gigapascals (15,000,000 psi), which is in agreement with quantum/atomistic models. Since carbon nanotubes have a low density for a solid of 1.3 to 1.4 g/cm^3, its specific strength of up to 48,000 kN·m·kg^{-1} is the best of known materials, compared to high-carbon steel's 154 kN·m·kg^{-1}.

Under excessive tensile strain, the tubes will undergo plastic deformation, which means the deformation is permanent. This deformation begins at strains of approximately 5% and can increase the maximum strain the tubes undergo before fracture by releasing strain energy.

Although the strength of individual CNT shells is extremely high, weak shear interactions between adjacent shells and tubes lead to significant reduction in the effective strength of multi-walled carbon nanotubes and carbon nanotube bundles down to only a few GPa. This limitation has been recently addressed by applying high-energy electron irradiation, which crosslinks inner shells and tubes, and effectively increases the strength of these materials to ≈60 GPa for multi-walled carbon nanotubes and ≈17 GPa for double-walled carbon nanotube bundles.

CNTs are not nearly as strong under compression. Because of their hollow structure and high aspect ratio, they tend to undergo buckling when placed under compressive, torsional, or bending stress.

Comparison of mechanical properties

Material	Young's modulus (TPa)	Tensile strength (GPa)	Elongation at break (%)
SWNT^E	≈1 (from 1 to 5)	13–53	16
Armchair SWNT^T	0.94	126.2	23.1
Zigzag SWNT^T	0.94	94.5	15.6–17.5
Chiral SWNT	0.92		
MWNT^E	0.2–0.8–0.95	11–63–150	
Stainless steel^E	0.186–0.214	0.38–1.55	15–50
Kevlar–29&149^E	0.06–0.18	3.6–3.8	≈2

^{E}Experimental observation; ^{T}Theoretical prediction

The above discussion referred to axial properties of the nanotube, whereas simple geometrical considerations suggest that carbon nanotubes should be much softer in the radial direction than along the tube axis. Indeed, TEM observation of radial elasticity suggested that even the van der Waals forces can deform two adjacent nanotubes. Nanoindentation experiments, performed by several groups on multiwalled carbon nanotubes and tapping/contact mode atomic force microscope measurements performed on single-walled carbon nanotubes, indicated a Young's modulus of the order of several GPa, confirming that CNTs are indeed rather soft in the radial direction.

Hardness

Standard single-walled carbon nanotubes can withstand a pressure up to 25 GPa without [plastic/permanent] deformation. They then undergo a transformation to superhard phase nanotubes. Maximum pressures measured using current experimental techniques are around 55 GPa. However, these new superhard phase nanotubes collapse at an even higher, albeit unknown, pressure.

The bulk modulus of superhard phase nanotubes is 462 to 546 GPa, even higher than that of diamond (420 GPa for single diamond crystal).

Wettability

The surface wettability of CNT is of importance for its applications in various settings. Although the intrinsic contact angle of graphite is around 90°, the contact angles of most as-synthesized CNT arrays are over 160°, exhibiting a superhydrophobic property. By applying a voltage as low as 1.3V, the extreme water repellant surface can be switched to a superhydrophilic one.

Kinetic properties

Multi-walled nanotubes are multiple concentric nanotubes precisely nested within one another. These exhibit a striking telescoping property whereby an inner nanotube core may slide, almost without friction, within its outer nanotube shell, thus creating an atomically perfect linear or rotational bearing. This is one of the first true examples of molecular nanotechnology, the precise positioning of atoms to create useful machines. Already, this property has been utilized to create the world's smallest rotational motor. Future applications such as a gigahertz mechanical oscillator are also envisioned.

Electrical Properties

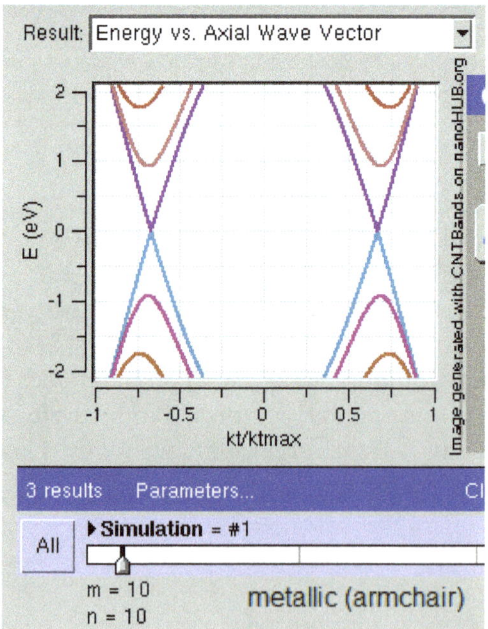

Band structures computed using tight binding approximation for (6,0) CNT (zigzag, metallic), (10,2) CNT (semiconducting) and (10,10) CNT (armchair, metallic).

Unlike graphene, which is a two-dimensional semimetal, carbon nanotubes are either metallic or semiconducting along the tubular axis. For a given (n,m) nanotube, if $n = m$, the nanotube is metallic; if $n - m$ is a multiple of 3, then the nanotube is semiconducting with a very small band gap, otherwise the nanotube is a moderate semiconductor. Thus all armchair ($n = m$) nanotubes are metallic, and nanotubes (6,4), (9,1), etc. are semiconducting. Carbon nanotubes are not semimetallic because the degenerate point (that point where the π [bonding] band meets the π^* [anti-bonding] band, at which the energy goes to zero) is slightly shifted away from the K point in the Brillouin zone due to the curvature of the tube surface, casing hybridization between the σ^* and π^* anti-bonding bands, modifying the band dispersion. The rule regarding metallic versus semiconductor behavior has exceptions, because curvature effects in small diameter tubes can strongly influence electrical properties. Thus, a (5,0) SWCNT that should be semiconducting in fact is metallic according to the calculations. Likewise, zigzag and chiral SWCNTs with small diameters that should be metallic have a finite gap (armchair nanotubes remain metallic). In theory, metallic nanotubes can carry an electric current density of 4×10^9 A/cm^2, which is more than 1,000 times greater than those of metals such as copper, where for copper interconnects current densities are limited by electromigration. Carbon nanotubes are thus being explored as conductivity enhancing components in composite materials and many groups are attempting to commercialize highly conducting electrical wire assembled from individual carbon nanotubes. There are significant challenges to be overcome, however, such as the much more resistive nanotube-to-nanotube junctions and impurities, all of which lower the electrical conductivity of the macroscopic nanotube wires by orders of magnitude, as compared to the conductivity of the individual nanotubes. Because of its nanoscale cross-section, electrons propagate only along the tube's axis. As a result, carbon nanotubes are frequently referred to as one-dimensional conductors. The maximum electrical conductance of a single-walled carbon nanotube is $2G_0$, where $G_0 = 2e^2/h$ is the conductance of a single ballistic quantum channel. Due to the role of the π-electron system in determining the electronic properties of graphene, doping in carbon nanotubes differs from that of bulk crystalline semiconductors from the same group of the periodic table (e.g. silicon). Graphitic substitution of carbon atoms in the nanotube wall by boron or nitrogen dopants leads to p-type and n-type behavior, respectively, as would be expected in silicon. However, some non-substitutional (intercalated or adsorbed) dopants introduced into a carbon nanotube, such as alkali metals as well as electron-rich metallocenes, result in n-type conduction because they donate electrons to the π-electron system of the nanotube. By contrast, π-electron acceptors such as $FeCl_3$ or electron-deficient metallocenes function as p-type dopants since they draw π-electrons away from the top of the valence band. Intrinsic superconductivity has been reported, although other experiments found no evidence of this, leaving the claim a subject of debate.

Optical properties

Thermal properties

Main article: Thermodynamics of nanostructures § Carbon nanotubes

All nanotubes are expected to be very good thermal conductors along the tube, exhibiting a property known as "ballistic conduction", but good insulators lateral to the tube axis. Measurements show that a SWNT has a room-temperature thermal conductivity along its axis of about 3500 $W \cdot m^{-1} \cdot K^{-1}$; compare this to copper, a metal well known for its good thermal conductivity, which transmits 385 $W \cdot m^{-1} \cdot K^{-1}$. A SWNT has a room-temperature thermal conductivity across its axis (in the radial direction) of about 1.52 $W \cdot m^{-1} \cdot K^{-1}$, which is about as thermally conductive as soil. The temperature stability of carbon nanotubes is estimated to be up to 2800 °C in vacuum and about 750 °C in air.

Defects

As with any material, the existence of a crystallographic defect affects the material properties. Defects can occur in the form of atomic vacancies. High levels of such defects can lower the tensile strength by up to 85%. An important example is the Stone Wales defect, which creates a pentagon and heptagon pair by rearrangement of the bonds. Because of the very small structure of CNTs, the tensile strength of the tube is dependent on its weakest segment in a similar manner to a chain, where the strength of the weakest link becomes the maximum strength of the chain. Crystallographic defects also affect the tube's electrical properties. A common result is lowered conductivity through the defective region of the tube. A defect in armchair-type tubes (which can conduct electricity) can cause the surrounding region to become semiconducting, and single monatomic vacancies induce magnetic properties.

Crystallographic defects strongly affect the tube's thermal properties. Such defects lead to phonon scattering, which in turn increases the relaxation rate of the phonons. This reduces the mean free path and reduces the thermal conductivity of nanotube structures. Phonon transport simulations indicate that substitutional defects such as nitrogen or boron will primarily lead to scattering of high-frequency optical phonons. However, larger-scale defects such as Stone Wales defects cause phonon scattering over a wide range of frequencies, leading to a greater reduction in thermal conductivity.

Safety and Health

The toxicity of carbon nanotubes has been an important question in nanotechnology. As of 2007, such research had just begun. The data is still fragmentary and subject to criticism. Preliminary results highlight the difficulties in evaluating the toxicity of this heterogeneous material. Parameters such as structure, size distribution, surface area, surface chemistry, surface charge, and agglomeration state as well as purity of the samples, have considerable impact on the reactivity of carbon nanotubes. However, available data clearly show that, under some conditions, nanotubes can cross membrane barriers, which suggests that, if raw materials reach the organs, they can induce harmful effects such as inflammatory and fibrotic reactions. As of October 2016, single wall carbon nanotubes have been registered through the European Union's Registration, Evaluation, Authorization and Restriction of Chemicals (REACH) regulations, based on evaluation of the potentially hazardous properties of SWCNT. Based on this registration, SWCNT commercialization is allowed in the EU up to 10 metric tons. Currently, the type of SWCNT registered through REACH is limited to the specific type of single wall carbon nanotubes manufactured by OCSiAl, which submitted the application.

Synthesis

Techniques have been developed to produce nanotubes in sizable quantities, including arc discharge, laser ablation, high-pressure carbon monoxide disproportionation, and chemical vapor deposition (CVD). Most of these processes take place in a vacuum or with process gases. CVD growth of CNTs can occur in vacuum or at atmospheric pressure. Large quantities of nanotubes can be synthesized by these methods; advances in catalysis and continuous growth are making CNTs more commercially viable.

Chemical modification

Carbon nanotubes can be functionalized to attain desired properties that can be used in a wide variety of applications. The two main methods of carbon nanotube functionalization are covalent and non-covalent modifications. Because of their hydrophobic nature, carbon nanotubes tend to agglomerate hindering their dispersion is solvents or viscous polymer melts. The resulting nanotube bundles or aggregates reduce the mechanical performance of the final composite. The surface of the carbon nanotubes can be modified to reduce the hydrophobicity and improve interfacial adhesion to a bulk polymer through chemical attachment.

Applications

Current

Current use and application of nanotubes has mostly been limited to the use of bulk nanotubes, which is a mass of rather unorganized fragments of nanotubes. Bulk nanotube materials may never achieve a tensile strength similar to that of individual tubes, but such composites may, nevertheless, yield strengths sufficient for many applications. Bulk carbon nanotubes have already been used as composite fibers in polymers to improve the mechanical, thermal and electrical properties of the bulk product.

- Easton-Bell Sports, Inc. have been in partnership with Zyvex Performance Materials, using CNT technology in a number of their bicycle components—including flat and riser handlebars, cranks, forks, seatposts, stems and aero bars.
- Zyvex Technologies has also built a 54' maritime vessel, the Piranha Unmanned Surface Vessel, as a technology demonstrator for what is possible using CNT technology. CNTs help improve the structural performance of the vessel, resulting in a lightweight 8,000 lb boat that can carry a payload of 15,000 lb over a range of 2,500 miles.
- Amroy Europe Oy manufactures Hybtonite carbon nanoepoxy resins where carbon nanotubes have been chemically activated to bond to epoxy, resulting in a composite material that is 20% to 30% stronger than other composite materials. It has been used for wind turbines, marine paints and variety of sports gear such as skis, ice hockey sticks, baseball bats, hunting arrows, and surfboards.

Other current applications include:

- tips for atomic force microscope probes
- in tissue engineering, carbon nanotubes can act as scaffolding for bone growth

There is also ongoing research in using carbon nanotubes as a scaffold for diverse microfabrication techniques.

Potential

The strength and flexibility of carbon nanotubes makes them of potential use in controlling other nanoscale structures, which suggests they will have an important role in nanotechnology engineering. The highest tensile strength of an individual multi-walled carbon nanotube has been tested to be 63 GPa. Carbon nanotubes were found in Damascus steel from the 17th century, possibly helping to account for the legendary strength of the swords made of it. Recently, several studies have highlighted the prospect of using carbon nanotubes as building blocks to fabricate three-dimensional macroscopic (>1mm in all three dimensions) all-carbon devices. Lalwani et al. have reported a novel radical initiated thermal crosslinking method to fabricated macroscopic, free-standing, porous, all-carbon scaffolds using single- and multi-walled carbon nanotubes as building blocks. These scaffolds possess macro-, micro-, and nano- structured pores and the porosity can be tailored for specific applications. These 3D all-carbon scaffolds/architectures maybe used for the fabrication of the next generation of energy storage, supercapacitors, field emission transistors, high-performance catalysis, photovoltaics, and biomedical devices and implants.

Discovery

The true identity of the discoverers of carbon nanotubes is a subject of some controversy. For years, scientists assumed that Sumio Iijima of NEC had discovered carbon nanotubes in 1991. He published a paper describing his discovery which initiated a flurry of excitement and could be credited by inspiring the many scientists now studying applications of carbon nanotubes. Though Iijima has been given much of the credit for discovering carbon nanotubes, it turns out that the timeline of carbon nanotubes goes back much further than 1991. In 1952 L. V. Radushkevich and V. M. Lukyanovich published clear images of 50 nanometer diameter tubes made of carbon in the Soviet *Journal of Physical Chemistry*. This discovery was largely unnoticed, as the article was published in Russian, and Western scientists' access to Soviet press was limited during the Cold War. Before they came to be known as carbon nanotubes, in 1976, Morinobu Endo of CNRS observed hollow tubes of rolled up graphite sheets synthesised by a chemical vapour-growth technique. The first specimens observed would later come to be known as single-walled carbon nanotubes (SWNTs). The three scientists have been the first ones to show images of a nanotube with a solitary graphene wall.

Endo, in his early review of vapor-phase-grown carbon fibers (VPCF), also reminded us that he had observed a hollow tube, linearly extended with parallel carbon layer faces near the fiber core. This appears to be the observation of multi-walled carbon nanotubes at the center of the fiber. The mass-produced MWCNTs today are strongly related to the VPGCF developed by Endo. In fact, they call it the "Endo-process", out of respect for his early work and patents.

In 1979, John Abrahamson presented evidence of carbon nanotubes at the 14th Biennial Conference of Carbon at Pennsylvania State University. The conference paper described carbon nanotubes as carbon fibers that were produced on carbon anodes during arc discharge. A characterization of these fibers was given as well as hypotheses for their growth in a nitrogen atmosphere at low pressures.

In 1981, a group of Soviet scientists published the results of chemical and structural characterization of carbon nanoparticles produced by a thermocatalytical disproportionation of carbon monoxide. Using TEM images and XRD patterns, the authors suggested that their "carbon multi-layer tubular crystals" were formed by rolling graphene layers into cylinders. They speculated that by rolling graphene layers into a cylinder, many different arrangements of graphene hexagonal nets are possible. They suggested two possibilities of such arrangements: circular arrangement (armchair nanotube) and a spiral, helical arrangement (chiral tube).

In 1987, Howard G. Tennent of Hyperion Catalysis was issued a U.S. patent for the production of "cylindrical discrete carbon fibrils" with a "constant diameter between about 3.5 and about 70 nanometers..., length 10^2 times the diameter, and an outer region of multiple essentially continuous layers of ordered carbon atoms and a distinct inner core...."

Iijima's discovery of multi-walled carbon nanotubes in the insoluble material of arc-burned graphite rods in 1991 and Mintmire, Dunlap, and White's independent prediction that if single-walled carbon nanotubes could be made, then they would exhibit remarkable conducting properties helped create the initial buzz that is now associated with carbon nanotubes. Nanotube research accelerated greatly following the independent discoveries by Bethune at IBM and Iijima at NEC of *single-walled* carbon nanotubes and methods to specifically produce them by adding transition-metal catalysts to the carbon in an arc discharge. The arc discharge technique was well-known to produce the famed Buckminster fullerene on a preparative scale, and these results appeared to extend the run of accidental discoveries relating to fullerenes. The discovery of nanotubes remains a contentious issue. Many believe that Iijima's report in 1991 is of particular importance because it brought carbon nanotubes into the awareness of the scientific community as a whole.

REFERENCES

1. Buzea, Cristina; Pacheco, Ivan; Robbie, Kevin (2007). "Nanomaterials and Nanoparticles: Sources and Toxicity". Biointerphases. **2** *(4): MR17–MR71. doi:10.1116/1.2815690. PMID 20419892.*
2. Eldridge, T. (8 January 2014). "Achieving industry integration with nanomaterials through financial markets". Nanotechnology_Now.
3. McGovern, C. (2010). "Commoditization of nanomaterials". Nanotechnol. Perceptions. **6** *(3): 155–178. doi:10.4024/N15GO10A.ntp.06.03.*
4. Novel natural nanomaterial spins off from spider-mite genome sequencing.

Phys.Org (May 23, 2013)

5. Why Are Tarantulas Blue?
6. «Fullerenes". Encyclopaedia Britannica.
7. Buseck, P.R.; Tsipursky, S.J.; Hettich, R. (1992). "Fullerenes from the Geological Environment". Science. **257** *(5067): 215–7. Bibcode:1992Sci...257..215B. doi:10.1126/science.257.5067.215. PMID 17794751.*
8. Cami, J; Bernard-Salas, J.; Peeters, E.; Malek, S. E. (2 September 2010). "Detection of C_{60} *and* C_{70} *in a Young Planetary Nebula". Science.* **329** *(5996): 1180–2. Bibcode:2010Sci...329.1180C. doi:10.1126/science.1192035. PMID 20651118.*
9. R. V. Lapshin (2016). "STM observation of a box-shaped graphene nanostructure appeared after mechanical cleavage of pyrolytic graphite" (PDF). Applied Surface Science. Netherlands: Elsevier B. V. **360**: 451–460. doi:10.1016/j.apsusc.2015.09.222. ISSN 0169-4332.
10. Zeng, S.; Baillargeat, Dominique; Ho, Ho-Pui; Yong, Ken-Tye (2014). "Nanomaterials enhanced surface plasmon resonance for biological and chemical sensing applications". Chemical Society Reviews. **43** *(10): 3426–3452. doi:10.1039/C3CS60479A. PMID 24549396.*
11. Kerativitayanan, P; Carrow, JK; Gaharwar, AK (26 May 2015). "Nanomaterials for Engineering Stem Cell Responses.". Advanced healthcare materials. **4**: 1600–27. doi:10.1002/adhm.201500272. PMID 26010739.
12. Wei, Hui; Wang, Erkang (2013-06-21). "Nanomaterials with enzyme-like characteristics (nanozymes): next-generation artificial enzymes". Chemical Society Reviews. **42** *(14). doi:10.1039/C3CS35486E. ISSN 1460-4744.*
13. Ramsden, J.J. (2011) Nanotechnology: An Introduction, Elsevier, Amsterdam
14. Zsigmondy, R. (1914) "Colloids and the Ultramicroscope", J. Wiley and Sons, NY
15. Dukhin, A.S. & Goetz, P.J. (2002). Ultrasound for characterizing colloids. Elsevier.
16. Onoda, G.Y. Jr.; Hench, L.L., eds. (1979). Ceramic Processing Before Firing. New York: Wiley & Sons. ISBN 0-471-65410-8.
17. Aksay, I.A.; Lange, F.F.; Davis, B.I. (1983). "Uniformity of Al_2O_3-ZrO_2Composites by Colloidal Filtration". J. Am. Ceram. Soc. **66** *(10): C–190. doi:10.1111/j.1151-2916.1983.tb10550.x.*
18. Franks, G.V. & Lange, F.F. (1996). "Plastic-to-Brittle Transition of Saturated, Alumina Powder Compacts". J. Am. Ceram. Soc. **79** *(12): 3161–3168. doi:10.1111/j.1151-2916.1996.tb08091.x.*
19. Evans, A.G.; Davidge, R.W. (1969). "The strength and fracture of fully dense polycrystalline magnesium oxide". Phil. Mag. **20** *(164): 373–388. Bibcode:1969PMag...20..373E. doi:10.1080/14786436908228708.*
20. Lange, F.F. & Metcalf, M. (1983). "Processing-Related Fracture Origins: II, Agglomerate Motion and Cracklike Internal Surfaces Caused by Differential Sintering". J. Am. Ceram. Soc. **66** *(6): 398–406. doi:10.1111/j.1151-2916.1983.tb10069.x.*
21. Evans, A.G. (1987). "Considerations of Inhomogeneity Effects in Sintering". J. Am. Ceram. Soc. **65** *(10): 497–501. doi:10.1111/j.1151-2916.1982.tb10340.x.*

22. Whitesides, George M.; et al. (1991). "Molecular Self-Assembly and Nanochemistry: A Chemical Strategy for the Synthesis of Nanostructures". Science. **254** *(5036): 1312–9. Bibcode:1991Sci...254.1312W. doi:10.1126/science.1962191. PMID 1962191.*

23. Dubbs D. M; Aksay I.A. (2000). "Self-Assembled Ceramics Produced by Complex-Fluid Templation". Annu. Rev. Phys. Chem. **51**: 601–22. Bibcode:2000ARPC...51..601D. doi:10.1146/annurev.physchem.51.1.601. PMID 11031294.

24. Nanomaterials. European Commission. Last updated 18 October 2011

25. ^ Jump up to:[a] [b] ISO/TS 800004-1 Nanotechnologies—Vocabulary—Part 1: Core terms. Geneva: 2011.

26. Kerativitayanan, Punyavee; Carrow, James K.; Gaharwar, Akhilesh K. (May 2015). "Nanomaterials for Engineering Stem Cell Responses". Advanced Healthcare Materials. **4**: n/a–n/a. doi:10.1002/adhm.201500272. PMID 26010739.

27. Lauterwasser, C. (18 July 2007). "Opportunities and risks of Nanotechnologies" (PDF). OECD.org.

28. Revell, P.A. (2006). "The biological effects of nanoparticles". Nanotechnology Perceptions. **2** *(3): 283–298. ISSN 1660-6795.*

29. «Study Sizes up Nanomaterial Toxicity". Chemical & Engineering News. **86** *(35): 44. 2008.*

30. Topmiller, Jennifer L.; Dunn, Kevin H. (9 December 2013). "Controlling Exposures to Workers Who Make or Use Nanomaterials". National Institute of Occupational Safety and Health. Retrieved 6 January 2015.

31. «Current Strategies for Engineering Controls in Nanomaterial Production and Downstream Handling Processes" (PDF). National Institute of Occupational Safety and Health. November 2013. Retrieved 6 January 2015.

32. «General Safe Practices for Working with Engineered Nanomaterials in Research Laboratories". National Institute of Occupational Safety and Health. May 2012. pp. 6–8. Retrieved 2016-07-15.

33. «Safety of Manufactured Nanomaterials: About, OECD Environment Directorate". OECD.org. 18 July 2007.

CHAPTER 2

Nanoparticle

Chapter Contents

A nanoparticle is an ultra-fine particle with at least one dimension between 1-100 nanometers (nm). One nanometer is equal to one billionth of a meter. The lower size limit helps to distinguish a particle from random clusters of atoms. The upper limit is the largest at which size related property differences normally manifest themselves.

This definition is widely accepted, though it is a bit arbitrary. There are published references to nanoparticles at sizes outside the 1-100 nm range. What makes such particles of interest to scientists are the unique material properties that sometimes result from their size. When particles manifest such properties, they will likely be considered nanoparticles even if they do not fit precisely within the defined size range.

INTRODUCTION

Nanoparticles are particles between 1 and 100 nanometers in size. In nanotechnology, a particle is defined as a small object that behaves as a whole unit with respect to its transport and properties. Particles are further classified according to diameter. Ultrafine particles are the same as nanoparticles and between 1 and 100 nanometers in size, fine particles are sized between 100 and 2,500 nanometers, and coarse particles cover a range between

2,500 and 10,000 nanometers. Scientific research on nanoparticles is intense as they have many potential applications in medicine, optics, and electronics. The U.S. National Nanotechnology Initiative offers government funding focused on nanoparticle research.

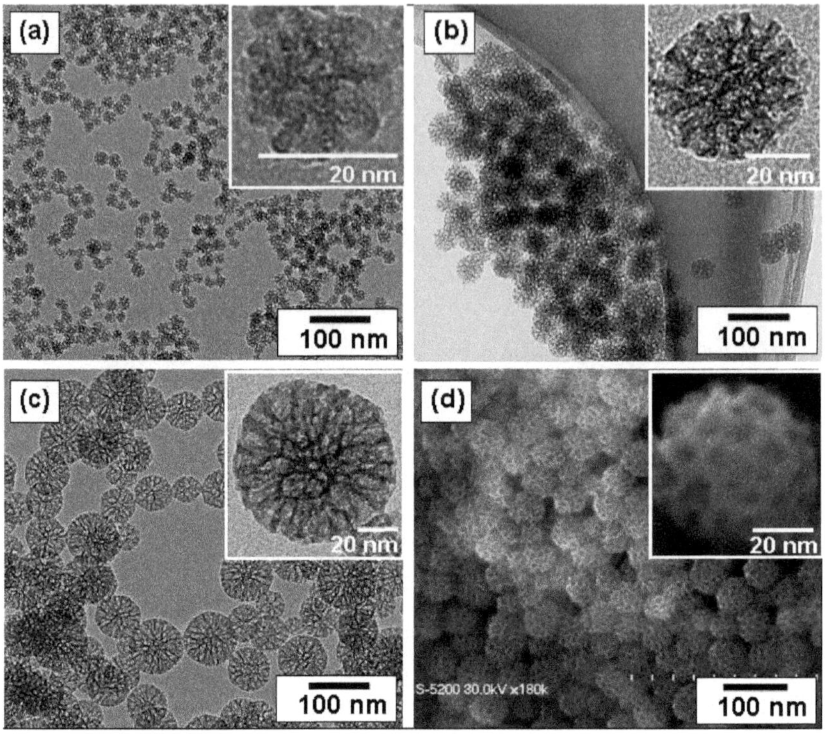

TEM (a, b, and c) images of prepared mesoporous silica nanoparticles with mean outer diameter: (a) 20nm, (b) 45nm, and (c) 80nm. SEM (d) image corresponding to (b). The insets are a high magnification of mesoporous silica particle.

DEFINITION

The reason for the synonymous definition of nanoparticles and ultrafine particles is that, during the 1970s and 80s, when the first thorough fundamental studies with "nanoparticles" were underway in the USA (by Granqvist and Buhrman)and Japan, (within an ERATO Project) they were called "ultrafine particles" (UFP). However, during the 1990s before the National Nanotechnology Initiative was launched in the USA, the new name, "nanoparticle," had become more common (for example, see the same senior author's paper 20 years later addressing the same issue, lognormal distribution of sizes). Nanoparticles can exhibit size-related properties significantly different from those of either fine particles or bulk materials.

Nanoclusters have at least one dimension between 1 and 10 nanometers and a narrow size distribution. Nanopowders are agglomerates of ultrafine particles, nanoparticles, or nanoclusters. Nanometer-sized single crystals, or single-domain ultrafine particles, are often referred to as nanocrystals.

BACKGROUND

Although nanoparticles are associated with modern science, they have a long history. Nanoparticles were used by artisans as far back as Rome in the fourth century in the famous Lycurgus cup made of dichroic glass as well as the ninth century in Mesopotamia for creating a glittering effect on the surface of pots.

In modern times, pottery from the Middle Ages and Renaissance often retains a distinct gold- or copper-colored metallic glitter. This luster is caused by a metallic film that was applied to the transparent surface of a glazing. The luster can still be visible if the film has resisted atmospheric oxidation and other weathering.

The luster originates within the film itself, which contains silver and copper nanoparticles dispersed homogeneously in the glassy matrix of the ceramic glaze. These nanoparticles are created by the artisans by adding copper and silver salts and oxides together with vinegar, ochre, and clay on the surface of previously-glazed pottery. The object is then placed into a kiln and heated to about 600 °C in a reducing atmosphere.

In the heat the glaze softens, causing the copper and silver ions to migrate into the outer layers of the glaze. There the reducing atmosphere reduced the ions back to metals, which then came together forming the nanoparticles that give the color and optical effects.

Luster technique showed that ancient craftsmen had a sophisticated empirical knowledge of materials. The technique originated in the Muslim world. As Muslims were not allowed to use gold in artistic representations, they sought a way to create a similar effect without using real gold. The solution they found was using luster.

Michael Faraday provided the first description, in scientific terms, of the optical properties of nanometer-scale metals in his classic 1857 paper. In a subsequent paper, the author (Turner) points out that: "It is well known that when thin leaves of gold or silver are mounted upon glass and heated to a temperature that is well below a red heat (~500 °C), a remarkable change of properties takes place, whereby the continuity of the metallic film is destroyed. The result is that white light is now freely transmitted, reflection is correspondingly diminished, while the electrical resistivity is enormously increased."

UNIFORMITY

The chemical processing and synthesis of high-performance technological components for the private, industrial, and military sectors requires the use of high-purity ceramics, polymers, glass-ceramics, and composite materials. In condensed bodies formed from fine powders, the irregular particle sizes and shapes in a typical powder often lead to non-uniform packing morphologies that result in packing density variations in the powder compact.

Uncontrolled agglomeration of powders due to attractive van der Waals forces can also give rise to microstructural heterogeneity. Differential stresses that develop as a result of non-uniform drying shrinkage are directly related to the rate at which the solvent can be removed, and thus highly dependent upon the distribution of porosity. Such stresses have been associated with a plastic-to-brittle transition in consolidated bodies, and can yield to crack propagation in the unfired body if not relieved.

In addition, any fluctuations in packing density in the compact as it is prepared for the kiln are often amplified during the sintering process, yielding inhomogeneous densification. Some pores and other structural defects associated with density variations have been shown to play a detrimental role in the sintering process by growing and thus limiting end-point densities. Differential stresses arising from inhomogeneous densification have also been shown to result in the propagation of internal cracks, thus becoming the strength-controlling flaws.

Inert gas evaporation and inert gas deposition are free many of these defects due to the distillation (cf. purification) nature of the process and having enough time to form single crystal particles, however even their non-aggreated deposits have lognormal size distribution, which is typical with nanoparticles. The reason why modern gas evaporation techniques can produce a relatively narrow size distribution is that aggregation can be avoided. However, even in this case, random residence times in the growth zone, due to the combination of drift and diffusion, result in a size distribution appearing lognormal.

It would, therefore, appear desirable to process a material in such a way that it is physically uniform with regard to the distribution of components and porosity, rather than using particle size distributions that will maximize the green density. The containment of a uniformly dispersed assembly of strongly interacting particles in suspension requires total control over interparticle forces. Monodisperse nanoparticles and colloids provide this potential.

Monodisperse powders of colloidal silica, for example, may therefore be stabilized sufficiently to ensure a high degree of order in the colloidal crystal or polycrystalline colloidal solid that results from aggregation. The degree of order appears to be limited by the time and space allowed for longer-range correlations to be established. Such defective polycrystalline colloidal structures would appear to be the basic elements of submicrometer colloidal materials science and, therefore, provide the first step in developing a more rigorous understanding of the mechanisms involved in microstructural evolution in high performance materials and components.

PROPERTIES

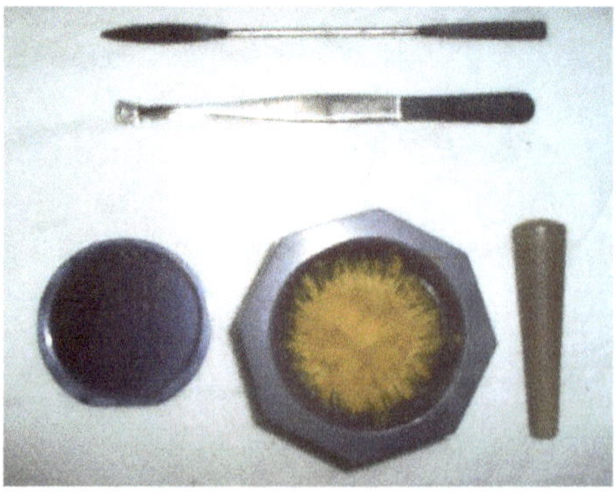

Silicon nanopowder

1 kg of particles of 1 mm^3 has the same surface area as 1 mg of particles of 1 nm^3.

Nanoparticles are of great scientific interest as they are, in effect, a bridge between bulk materials and atomic or molecular structures. A bulk material should have constant physical properties regardless of its size, but at the nano-scale size-dependent properties are often observed. Thus, the properties of materials change as their size approaches the nanoscale and as the percentage of the surface in relation to the percentage of the volume of a material becomes significant. For bulk materials larger than one micrometer (or micron), the percentage of the surface is insignificant in relation to the volume in the bulk of the material. The interesting and sometimes unexpected properties of nanoparticles are therefore largely due to the large surface area of the material, which dominates the contributions made by the small bulk of the material.

Nanoparticles often possess unexpected optical properties as they are small enough to confine their electrons and produce quantum effects. For example, gold nanoparticles appear deep-red to black in solution. Nanoparticles of yellow gold and grey silicon are red in color. Gold nanoparticles melt at much lower temperatures (~300 °C for 2.5 nm size) than the gold slabs (1064 °C);. Absorption of solar radiation is much higher in materials composed of nanoparticles than it is in thin films of continuous sheets of material. In both solar PV and solar thermal applications, controlling the size, shape, and material of the particles, it is possible to control solar absorption.

Other size-dependent property changes include quantum confinement in semiconductor particles, surface plasmon resonance in some metal particles and superparamagnetism in magnetic materials. What would appear ironic is that the changes in physical properties are not always desirable. Ferromagnetic materials smaller than 10 nm can switch their magnetisation direction using room temperature thermal energy, thus making them unsuitable for memory storage.

Suspensions of nanoparticles are possible since the interaction of the particle surface with the solvent is strong enough to overcome density differences, which otherwise usually result in a material either sinking or floating in a liquid.

The high surface area to volume ratio of nanoparticles provides a tremendous driving force for diffusion, especially at elevated temperatures. Sintering can take place at lower temperatures, over shorter time scales than for larger particles. In theory, this does not affect the density of the final product, though flow difficulties and the tendency of nanoparticles to agglomerate complicates matters. Moreover, nanoparticles have been found to impart some extra properties to various day to day products. For example, the presence of titanium dioxide nanoparticles imparts what we call the self-cleaning effect, and, the size being nano-range, the particles cannot be observed. Zinc oxide particles have been found to have superior UV blocking properties compared to its bulk substitute. This is one of the reasons why it is often used in the preparation of sunscreen lotions, is completely photostable and toxic.

Clay nanoparticles when incorporated into polymer matrices increase reinforcement, leading to stronger plastics, verifiable by a higher glass transition temperature and other mechanical property tests. These nanoparticles are hard, and impart their properties to the polymer (plastic). Nanoparticles have also been attached to textile fibers in order to create smart and functional clothing.

Metal, dielectric, and semiconductor nanoparticles have been formed, as well as hybrid structures (e.g., core–shell nanoparticles). Nanoparticles made of semiconducting material may also be labeled quantum dots if they are small enough (typically sub 10 nm) that quantization of electronic energy levels occurs. Such nanoscale particles are used in biomedical applications as drug carriers or imaging agents.

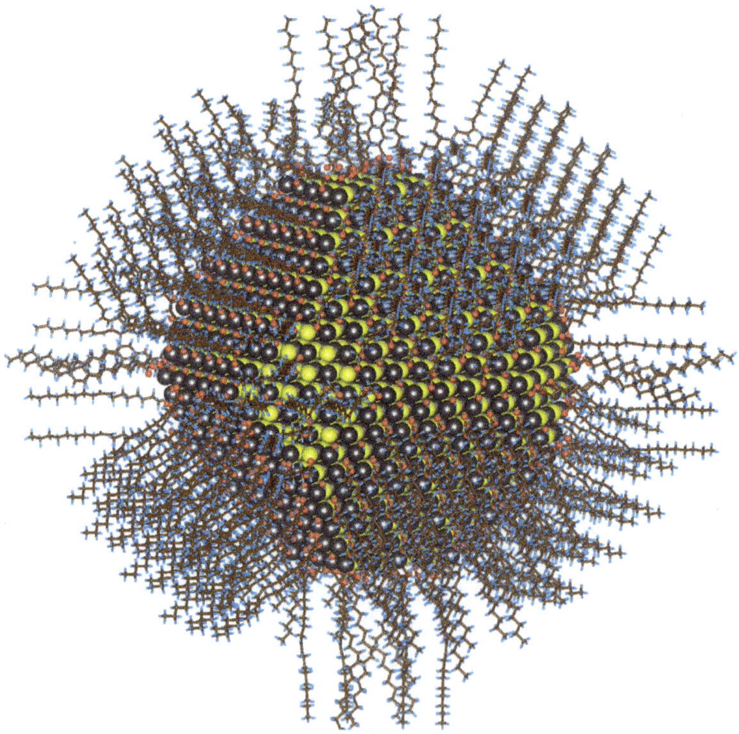

Semiconductor nanoparticle (quantum dot) of lead sulfide with complete passivation by oleic acid, oleyl amine and hydroxyl ligands (size ~5nm).

Semi-solid and soft nanoparticles have been manufactured. A prototype nanoparticle of semi-solid nature is the liposome. Various types of liposome nanoparticles are currently used clinically as delivery systems for anticancer drugs and vaccines.

Nanoparticles with one half hydrophilic and the other half hydrophobic are termed Janus particles and are particularly effective for stabilizing emulsions. They can self-assemble at water/oil interfaces and act as solid surfactants.

SYNTHESIS

There are several methods for creating nanoparticles, including attrition, pyrolysis and hydrothermal synthesis. In attrition, macro- or micro-scale particles are ground in a ball mill, a planetary ball mill, or other size-reducing mechanism. The resulting particles are air classified to recover nanoparticles. In pyrolysis, a vaporous precursor (liquid or gas) is forced through an orifice at high pressure and burned. The resulting solid (a version of soot) is air classified to recover oxide particles from by-product gases. Traditional pyrolysis often results in aggregates and agglomerates rather than single primary particles. Ultrasonic nozzle spray pyrolysis (USP) on the other hand aids in preventing agglomerates from forming.

A thermal plasma can deliver the energy to vaporize small micrometer-size particles. The thermal plasma temperatures are in the order of 10,000 K, so that solid powder easily evaporates. Nanoparticles are formed upon cooling while exiting the plasma region. The main types of the thermal plasma torches used to produce nanoparticles are dc plasma jet, dc arc plasma, and radio frequency (RF) induction plasmas. In the arc plasma reactors, the energy necessary for evaporation and reaction is provided by an electric arc formed between the anode and the cathode. For example, silica sand can be vaporized with an arc plasma at atmospheric pressure, or thin aluminum wires can be vaporized by exploding wire method. The resulting mixture of plasma gas and silica vapour can be rapidly cooled by quenching with oxygen, thus ensuring the quality of the fumed silica produced.

In RF induction plasma torches, energy coupling to the plasma is accomplished through the electromagnetic field generated by the induction coil. The plasma gas does not come in contact with electrodes, thus eliminating possible sources of contamination and allowing the operation of such plasma torches with a wide range of gases including inert, reducing, oxidizing, and other corrosive atmospheres. The working frequency is typically between 200 kHz and 40 MHz. Laboratory units run at power levels in the order of 30–50 kW, whereas the large-scale industrial units have been tested at power levels up to 1 MW. As the residence time of the injected feed droplets in the plasma is very short, it is important that the droplet sizes are small enough in order to obtain complete evaporation. The RF plasma method has been used to synthesize different nanoparticle materials, for example synthesis of various ceramic nanoparticles such as oxides, carbours/carbides, and nitrides of Ti and Si (see Induction plasma technology).

Inert-gas condensation is frequently used to make nanoparticles from metals with low melting points. The metal is vaporized in a vacuum chamber and then supercooled with an inert gas stream. The supercooled metal vapor condenses into nanometer-size particles, which can be entrained in the inert gas stream and deposited on a substrate or studied in situ.

Nanoparticles can also be formed using radiation chemistry. Radiolysis from gamma rays can create strongly active free radicals in solution. This relatively simple technique uses a minimum number of chemicals. These including water, a soluble metallic salt, a radical scavenger (often a secondary alcohol), and a surfactant (organic capping agent). High gamma doses on the order of 10^4 Gray are required. In this process, reducing radicals will drop metallic ions down to the zero-valence state. A scavenger chemical will preferentially interact with oxidizing radicals to prevent the re-oxidation of the metal. Once in the zero-valence state, metal atoms begin to coalesce into particles. A chemical surfactant surrounds the particle during formation and regulates its growth. In sufficient concentrations, the surfactant molecules stay attached to the particle. This prevents it from dissociating or forming clusters with other particles. Formation of nanoparticles using the radiolysis method allows for tailoring of particle size and shape by adjusting precursor concentrations and gamma dose.

Sol-gel

The sol-gel process is a wet-chemical technique (also known as chemical solution deposition) widely used recently in the fields of materials science and ceramic engineering. Such methods are used primarily for the fabrication of materials (typically a metal oxide) starting from a chemical solution (sol, short for solution), which acts as the precursor for an integrated network (or gel) of either discrete particles or network polymers.

Typical precursors are metal alkoxides and metal chlorides, which undergo hydrolysis and polycondensation reactions to form either a network "elastic solid" or a colloidal suspension (or dispersion) – a system composed of discrete (often amorphous) submicrometer particles dispersed to various degrees in a host fluid. Formation of a metal oxide involves connecting the metal centers with oxo (M-O-M) or hydroxo (M-OH-M) bridges, therefore generating metal-oxo or metal-hydroxo polymers in solution. Thus, the sol evolves toward the formation of a gel-like diphasic system containing both a liquid phase and solid phase whose morphologies range from discrete particles to continuous polymer networks.

In the case of the colloid, the volume fraction of particles (or particle density) may be so low that a significant amount of fluid may need to be removed initially for the gel-like properties to be recognized. This can be accomplished in a number of ways. The most simple method is to allow time for sedimentation to occur, and then pour off the remaining liquid. Centrifugation can also be used to accelerate the process of phase separation.

Removal of the remaining liquid (solvent) phase requires a drying process, which is typically causes shrinkage and densification. The rate at which the solvent can be removed is ultimately determined by the distribution of porosity in the gel. The ultimate microstructure of the final component will clearly be strongly influenced by changes implemented during this phase of processing. Afterward, a thermal treatment, or firing process, is often necessary in order to favor further polycondensation and enhance mechanical properties and structural stability via final sintering, densification, and grain growth. One of the distinct advantages of using this methodology as opposed to the more traditional processing techniques is that densification is often achieved at a much lower temperature.

The precursor sol can be either deposited on a substrate to form a film (e.g., by dip-coating or spin-coating), cast into a suitable container with the desired shape (e.g., to obtain a monolithic ceramics, glasses, fibers, membranes, aerogels), or used to synthesize powders (e.g., microspheres, nanospheres). The sol-gel approach is a cheap and low-temperature technique that allows for the fine control of the product's chemical composition. Even small quantities of dopants, such as organic dyes and rare earth metals, can be introduced in the sol and end up uniformly dispersed in the final product. It can be used in ceramics processing and manufacturing as an investment casting material, or as a means of producing very thin films of metal oxides for various purposes. Sol-gel derived materials have diverse applications in optics, electronics, energy, space, (bio) sensors, medicine (e.g., controlled drug release) and separation (e.g., chromatography) technology.

COLLOIDS

The term colloid is used primarily to describe a range of mixtures which have solid particles dispersed in a liquid medium. The term applies only if the particles are larger than atomic dimensions but small enough to exhibit Brownian motion. If the particles are large enough, their dynamic behavior in any given period of time in suspension would be governed by forces of gravity and sedimentation. If the particles are small enough, their irregular motion in suspension can be attributed to the collective bombardment of a myriad of thermally agitated molecules in the liquid suspending medium, as described originally by Albert Einstein in his dissertation. Einstein showed evidence that water was made up of discrete molecules by characterizing the observed erratic particle behavior using the theory of Brownian motion, with sedimentation being a possible result. The critical size range (or particle diameter) typically ranges from nanometers (10^{-9} m) to micrometers (10^{-6} m).

MORPHOLOGY

Nanostars of vanadium(IV) oxide.

Scientists have taken to naming their particles after the real-world shapes that they might represent. Nanospheres, nanochains, nanoreefs, nanoboxes and more have appeared in the literature. These morphologies sometimes arise spontaneously as an effect of a templating or directing agent present in the synthesis such as miscellar emulsions or anodized alumina pores, or from the innate crystallographic growth patterns of the materials themselves. Some of these morphologies may serve a purpose, such as long carbon nanotubes used to bridge an electrical junction, or just a scientific curiosity like the stars shown at right.

Amorphous particles usually adopt a spherical shape (due to their microstructural isotropy), whereas the shape of anisotropic microcrystalline whiskers corresponds to their particular crystal habit. At the small end of the size range, nanoparticles are often referred to as clusters. Spheres, rods, fibers, and cups are just a few of the shapes that have been grown. The study of fine particles is called micromeritics.

CHARACTERIZATION

The majority of nanoparticle characterization techniques are light-based, but a non-optical nanoparticle characterization technique called Tunable Resistive Pulse Sensing (TRPS) has been developed that enables the simultaneous measurement of size, concentration and surface charge for a wide variety of nanoparticles. This technique, which applies the Coulter Principle, allows for particle-by-particle quantification of these three nanoparticle characteristics with high resolution.

FUNCTIONALIZATION

The surface coating of nanoparticles determines many of their properties, notably stability, solubility, and targeting. A coating that is multivalent or polymeric confers high stability. Functionalized nanomaterial-based catalysts can be used for catalysis of many known organic reactions.

Surface coating for Biological Applications

For biological applications, the surface coating should be polar to give high aqueous solubility and prevent nanoparticle aggregation. In serum or on the cell surface, highly charged coatings promote non-specific binding, whereas polyethylene glycol linked to terminal hydroxyl or methoxy groups repel non-specific interactions. Nanoparticles can be linked to biological molecules that can act as address tags, to direct the nanoparticles to specific sites within the body, specific organelles within the cell, or to follow specifically the movement of individual protein or RNA molecules in living cells. Common address tags are monoclonal antibodies, aptamers, streptavidin or peptides. These targeting agents should ideally be covalently linked to the nanoparticle and should be present in a controlled number per nanoparticle. Multivalent nanoparticles, bearing multiple targeting groups, can cluster receptors, which can activate cellular signaling pathways, and give stronger anchoring. Monovalent nanoparticles, bearing a single binding site, avoid clustering and so are preferable for tracking the behavior of individual proteins.

Red blood cell coatings can help nanoparticles evade the immune system.

SAFETY

Nanoparticles present possible dangers, both medically and environmentally. Most of these are due to the high surface to volume ratio, which can make the particles very reactive or catalytic. They are also able to pass through cell membranes in organisms, and their interactions with biological systems are relatively unknown. However, it is unlikely the particles would enter the cell nucleus, Golgi complex, endoplasmic reticulum or other internal cellular components due to the particle size and intercellular agglomeration. A recent study looking at the effects of ZnO nanoparticles on human immune cells has found varying levels of susceptibility to cytotoxicity. There are concerns that pharmaceutical companies, seeking regulatory approval for nano-reformulations of existing medicines, are relying on safety data produced during clinical studies of the earlier, pre-reformulation version of the medicine. This could result in regulatory bodies, such as the FDA, missing new side effects that are specific to the nano-reformulation. Whether cosmetics and sunscreens containing nanomaterials pose health risks remains largely unknown at this stage. However considerable research has demonstrated that zinc nanoparticles are not absorbed into the bloodstream in vivo. Concern has also been raised over the health effects of respirable nanoparticles from certain combustion processes. As of 2013 the U.S. Environmental Protection Agency was investigating the safety of the following nanoparticles:

- Carbon Nanotubes: Carbon materials have a wide range of uses, ranging from composites for use in vehicles and sports equipment to integrated circuits for electronic components. The interactions between nanomaterials such as carbon nanotubes and natural organic matter strongly influence both their aggregation and deposition, which strongly affects their transport, transformation, and exposure in aquatic environments. In past research, carbon nanotubes exhibited some toxicological impacts that will be evaluated in various environmental settings in current EPA chemical safety research. EPA research will provide data, models, test methods, and best practices to discover the acute health effects of carbon nanotubes and identify methods to predict them.
- Cerium oxide: Nanoscale cerium oxide is used in electronics, biomedical supplies, energy, and fuel additives. Many applications of engineered cerium oxide nanoparticles naturally disperse themselves into the environment, which increases the risk of exposure. There is ongoing exposure to new diesel emissions using fuel additives containing CeO_2 nanoparticles, and the environmental and public health impacts of this new technology are unknown. EPA's chemical safety research is assessing the environmental, ecological, and health implications of nanotechnology-enabled diesel fuel additives.
- Titanium dioxide: Nano titanium dioxide is currently used in many products. Depending on the type of particle, it may be found in sunscreens, cosmetics, and paints and coatings. It is also being investigated for use in removing contaminants from drinking water.
- Nano Silver: Nano silver is being incorporated into textiles, clothing, food packaging, and other materials to eliminate bacteria. EPA and the U.S. Consumer Product Safety Commission are studying certain products to see whether they

transfer nano-size silver particles in real-world scenarios. EPA is researching this topic to better understand how much nano-silver children come in contact with in their environments.

- Iron: While nano-scale iron is being investigated for many uses, including "smart fluids" for uses such as optics polishing and as a better-absorbed iron nutrient supplement, one of its more prominent current uses is to remove contamination from groundwater. This use, supported by EPA research, is being piloted at a number of sites across the country.

LASER APPLICATIONS

The use of nanoparticles in laser dye-doped poly(methyl methacrylate) (PMMA) laser gain media was demonstrated in 2003 and it has been shown to improve conversion efficiencies and to decrease laser beam divergence. Researchers attribute the reduction in beam divergence to improved dn/dT characteristics of the organic-inorganic dye-doped nanocomposite. The optimum composition reported by these researchers is 30% w/w of SiO_2 (~ 12 nm) in dye-doped PMMA.

MEDICINAL APPLICATIONS

Liposome

Though liposomes can vary in size from low micrometer range to tens of micrometers, unilamellar liposomes, as pictured here, are typically in the lower size range with various targeting ligands attached to their surface allowing for their surface-attachment and accumulation in pathological areas for treatment of disease.

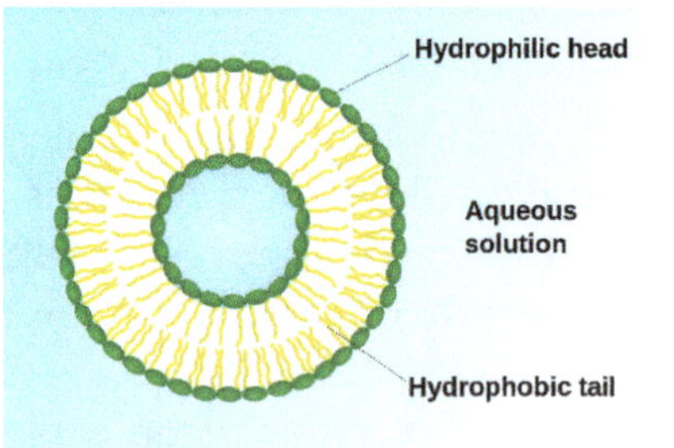

Scheme of a liposome formed by phospholipids in an aqueous solution.

A **liposome** is a spherical vesicle having at least one lipid bilayer. The liposome can be used as a vehicle for administration of nutrients and pharmaceutical drugs. Liposomes can be prepared by disrupting biological membranes (such as by sonication). Liposomes are most often composed of phospholipids, especially phosphatidylcholine, but may also include other lipids, such as eggphosphatidylethanolamine, so long as they are compatible with lipid bilayer structure. A liposome design may employ surface ligands for attaching to unhealthy tissue. The major types of liposomes are the multilamellar vesicle (MLV, with several lamellar phase lipid bilayers), the small unilamellar liposome vesicle (SUV, with one lipid bilayer), the large unilamellar vesicle (LUV), and the cochleate vesicle. A less desirable form are multivesicular liposomes in which one vesicle contains one or more smaller vesicles.

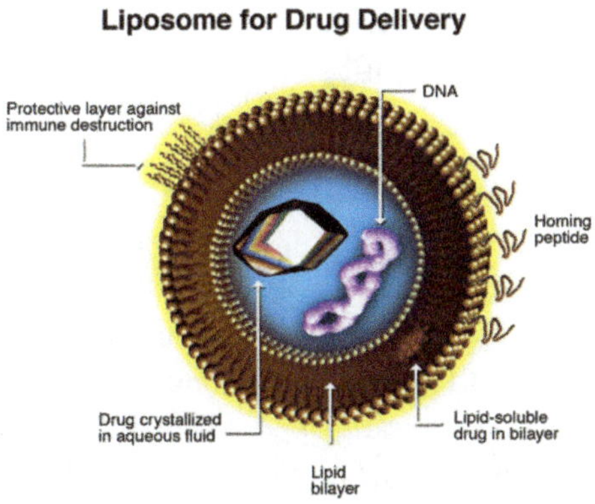

Liposomes are composite structures made of phospholipids and may contain small amounts of other molecules.

Liposomes should not be confused with micelles and reverse micelles composed of monolayers.

Discovery

The word liposome derives from two Greek words: lipo ("fat") and soma ("body"); it is so named because its composition is primarily of phospholipid. Liposomes were first described by British haematologist Alec D Bangham in 1961 (published 1964), at the Babraham Institute, in Cambridge. They were discovered when Bangham and R. W. Horne were testing the institute's new electron microscope by adding negative stain to dry phospholipids. The resemblance to the plasmalemma was obvious, and the microscope pictures served as the first evidence for the cell membrane being a bilayer lipid structure. Their integrity as a closed, bilayer structure, that could release its contents after detergent treatment (structure-linked latency) was established by Bangham, Standish and Weissmann in the next year. Weissmann - during a Cambridge pub discussion with Bangham - first named the structures "liposomes" after the lysosome, which his laboratory had been studying: a simple organelle the structure-linked latency of which could be disrupted by detergents and streptolysins. Liposomes can be easily distinguished from micelles and hexagonal lipid phases by negative staining transmission electron microscopy. Alec Douglas Bangham with colleagues Jeff Watkins and Malcolm Standish wrote the 1965 paper that effectively launched the liposome "industry". Around this time he was joined at Babraham by Gerald Weissmann, an American physician with an interest in lysosomes. Now an emeritus professor at New York University School of Medicine, Weissmann recalls the two of them sitting in a Cambridge pub and reflecting on the role of lipid sheets in separating the interior of the cell from the exterior milieu. This insight, they felt, was to cell function what the discovery of the double helix had been to genetics. Bangham had called his lipid structures "multilamellar smectic mesophases" or sometimes "Banghasomes". It was Weissmann who proposed the more user-friendly term liposome.

Mechanism

A liposome has an aqueous solution core surrounded by a hydrophobic membrane, in the form of a lipid bilayer; hydrophilic solutes dissolved in the core cannot readily pass through the bilayer. Hydrophobic chemicals associate with the bilayer. A liposome can be hence loaded with hydrophobic and/or hydrophilic molecules. To deliver the molecules to a site of action, the lipid bilayer can fuse with other bilayers such as the cell membrane, thus delivering the liposome contents; this is a complex and non-spontaneous event, however. By preparing liposomes in a solution of DNA or drugs (which would normally be unable to diffuse through the membrane) they can be (indiscriminately) delivered past the lipid bilayer, but are then typically distributed non-homogeneously. Liposomes are used as models for artificial cells. Liposomes can also be designed to deliver drugs in other ways. Liposomes that contain low (or high) pH can be constructed such that dissolved aqueous drugs will be charged in solution (i.e., the pH is outside the drug's pI range). As the pH naturally neutralizes within the liposome (protons can pass through some membranes), the drug will also be neutralized, allowing it to freely pass through a membrane. These liposomes work to deliver drug by diffusion rather than by direct cell fusion. A similar approach can be exploited in the biodetoxification of drugs by injecting empty liposomes with a transmembrane pH gradient. In this case the vesicles act as sinks to scavenge the drug in the blood circulation and prevent its toxic effect.Another strategy for liposome drug delivery is to target endocytosis events. Liposomes can be made in a particular size range that makes them viable targets for natural macrophage phagocytosis. These liposomes may be digested while in the macrophage's phagosome, thus releasing its drug. Liposomes can also be decorated with opsonins and ligands to activate endocytosis in other cell types.

The use of liposomes for transformation or transfection of DNA into a host cell is known as lipofection.

In addition to gene and drug delivery applications, liposomes can be used as carriers for the delivery of dyes to textiles, pesticides to plants, enzymes and nutritional supplements to foods, and cosmetics to the skin.

Liposomes are also used as outer shells of some microbubble contrast agents used in contrast-enhanced ultrasound.

List of Drugs

As of 2012, 13 drugs with liposomal delivery systems have been approved and five additional liposomal drugs were in clinical trials.

List of Clinically Approved

Name	Trade name	Company	**Indication**	**Liposomal Excipients**
Liposomal amphotericin B	Abelcet	Enzon	Fungal infections	DMPC and DMPG
Liposomal amphotericin B	Ambisome	Gilead Sciences	Fungal and protozoal infections	HSPC, Cholesterol, DSPG

Liposomal cytarabine	Depocyt	Pacira (formerly SkyePharma)	Malignant lymphomatous meningitis	DOPC, Cholesterol, DPPG
Liposomal daunorubicin	DaunoXome	Gilead Sciences	HIV-related Kaposi's sarcoma	DSPC, Cholesterol
Liposomal doxorubicin	Myocet	Zeneus	Combination therapy with cyclophosphamide in metastatic breast cancer	LIPOVA-E120, Cholesterol
Liposomal IRIV vaccine	Epaxal	Crucell	Hepatitis A	LECIVA-S70
Liposomal IRIV vaccine	Inflexal V	Berna Biotech	Influenza	LECIVA-S90
Liposomal morphine	DepoDur	SkyePharma, Endo	Postsurgical analgesia	DOPC, Cholesterol, DPPG
Liposomal verteporfin	Visudyne	QLT, Novartis	Age-related macular degeneration, pathologic myopia, ocular histoplasmosis	Egg PG, DMPC
Liposome-proteins SP-B and SP-C	Curosurf	Chiesi Farmaceutici, S.p.A.	pulmonary surfactant for Respiratory Distress Syndrome (RDS)	Leciva-S90
Liposome-PEG doxorubicin	Doxil/ Caelyx	Ortho Biotech, Schering-Plough	HIV-related Kaposi's sarcoma, metastatic breast cancer, metastatic ovarian cancer	MPEG-DSPE, HSPC, Cholesterol
Micellular estradiol	Estrasorb	Novavax	Menopausal therapy	Soybean oil, Polysorbate80
Liposomal vincristine	Marqibo	Spectrum Pharmaceuticals	Acute Lymphoblastic Leukemia (ALL) and Melanoma	Cholesterol and egg sphingomyelin
Liposome-PEG doxorubicin	Lipo-Dox	TTY BIOPHARM	HIV-related Kaposi's sarcoma, metastatic breast cancer, metastatic ovarian cancer, Multiple Myeloma	MPEG-DSPE, DSPC, Cholesterol

Dietary and Nutritional Supplements

Regarding the use of liposomes as a carrier of dietary and nutritional supplements; until very recently the use of liposomes were primarily directed at targeted drug delivery. However, the versatile abilities of liposomes are now being discovered in other settings. Liposomes are presently being implemented for the specific oral delivery of certain dietary and nutritional supplements.

A very small number of dietary and nutritional supplement companies are currently pioneering the benefits of this unique science towards this new application. This new direction and employment of liposome science is in part due to the low absorption and bioavailability rates of traditional oral dietary and nutritional tablets and capsules. The low oral bioavailability and absorption of many nutrients is clinically well documented. Therefore, the natural encapsulation of lypophilic and hydrophilic nutrients within liposomes has made for a very effective method of bypassing the destructive elements of the gastric system and aiding the encapsulated nutrient to be delivered to the cells and tissues. It is important to note that certain influential factors have far reaching effects on the percentage of liposome that are yielded in manufacturing. These influences also have an effect on the actual amount of realized liposome entrapment and the actual quality of the liposomes themselves. These are very crucial elements which lead to the long term stability of the liposomes. These complex yet significant factors are the following: (1) The actual manufacturing method and preparation of the liposomes themselves; (2) The constitution, quality, and type of raw phospholipid used in the formulation and manufacturing of the liposomes; (3) The ability to create homogeneous liposome particle sizes that are stable and hold their encapsulated payload. These primary and key elements comprise the foundation of an effective liposome carrier for use in increasing the bioavailability of oral dosages of dietary and nutritional supplements.

Manufacturing

The choice of liposome preparation method depends, i.a., on the following parameters:

- the physicochemical characteristics of the material to be entrapped and those of the liposomal ingredients;
- the nature of the medium in which the lipid vesicles are dispersed
- the effective concentration of the entrapped substance and its potential toxicity;
- additional processes involved during application/delivery of the vesicles;
- optimum size, polydispersity and shelf-life of the vesicles for the intended application; and,
- batch-to-batch reproducibility and possibility of large-scale production of safe and efficient liposomal products

Useful liposomes rarely form spontaneously. They typically form after supplying enough energy to a dispersion of (phospho)lipids in a polar solvent, such as water, to break down multilamellar aggregates into oligo- or unilamellar bilayer vesicles. Liposomes can hence be created by sonicating a dispersion of amphipatic lipids, such as phospholipids, in water. Low shear rates create multilamellar liposomes. The original aggregates, which have many layers like an onion, thereby form progressively smaller and finally unilamellar liposomes (which are often unstable, owing to their small size and the sonication-created defects). Sonication is generally considered a “gross” method of preparation as it can damage the structure of the drug to be encapsulated. Newer methods such as extrusion and Mozafari method are employed to produce materials for human use. Using lipids other than phosphatidylcholine can greatly facilitate liposome preparation.

Prospect

Further advances in liposome research have been able to allow liposomes to avoid detection by the body's immune system, specifically, the cells of reticuloendothelial system (RES). These liposomes are known as "stealth liposomes". They were first proposed by G. Cevc and G. Blume and, independently and soon thereafter, the groups of L. Huang and V. Torchilin and are constructed with PEG (Polyethylene Glycol) studding the outside of the membrane. The PEG coating, which is inert in the body, allows for longer circulatory life for the drug delivery mechanism. However, research currently seeks to investigate at what amount of PEG coating the PEG actually hinders binding of the liposome to the delivery site. In addition to a PEG coating, most stealth liposomes also have some sort of biological species attached as a ligand to the liposome, to enable binding via a specific expression on the targeted drug delivery site. These targeting ligands could be monoclonal antibodies (making an immunoliposome), vitamins, or specific antigens, but must be accessible. Targeted liposomes can target nearly any cell type in the body and deliver drugs that would otherwise be systemically delivered. Naturally toxic drugs can be much less systemically toxic if delivered only to diseased tissues. Polymersomes, morphologically related to liposomes, can also be used this way. Also morphologically related to liposomes are highly deformable vesicles, designed for non-invasive transdermal material delivery, known as Transfersomes.

Certain anticancer drugs such as doxorubicin (Doxil) and daunorubicin may be administered via liposomes. Liposomal cisplatin has received orphan drug designation for pancreatic cancer from EMEA.

DENDRIMER

Dendrimers are repetitively branchedmolecules. The name comes from the Greek word δέντρο (**dendro**). which translates to "tree".The Greek word δενδριμερές is translated as a part of the tree. It is named like that, because it is like to see a tree from the top. Synonymous terms for dendrimer include **arborols** and **cascade molecules**. However, dendrimer is currently the internationally accepted term. A dendrimer is typically symmetric around the core, and often adopts a spherical three-dimensional morphology. The word **dendron** is also encountered frequently. A dendron usually contains a single chemically addressable group called the focal point or core. The difference between dendrons and dendrimers is illustrated in the top figure, but the terms are typically encountered interchangeably.

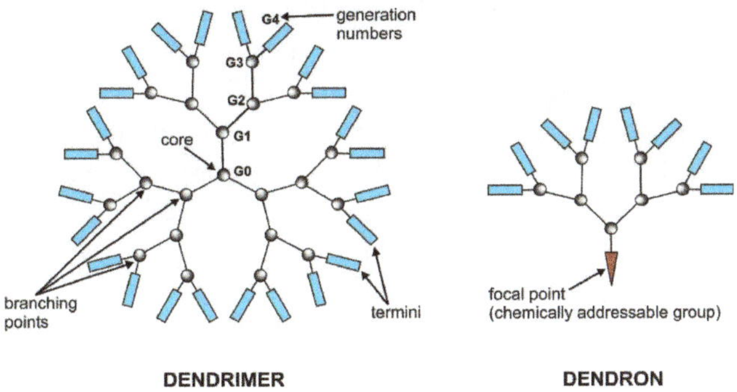

Dendrimer and dendron

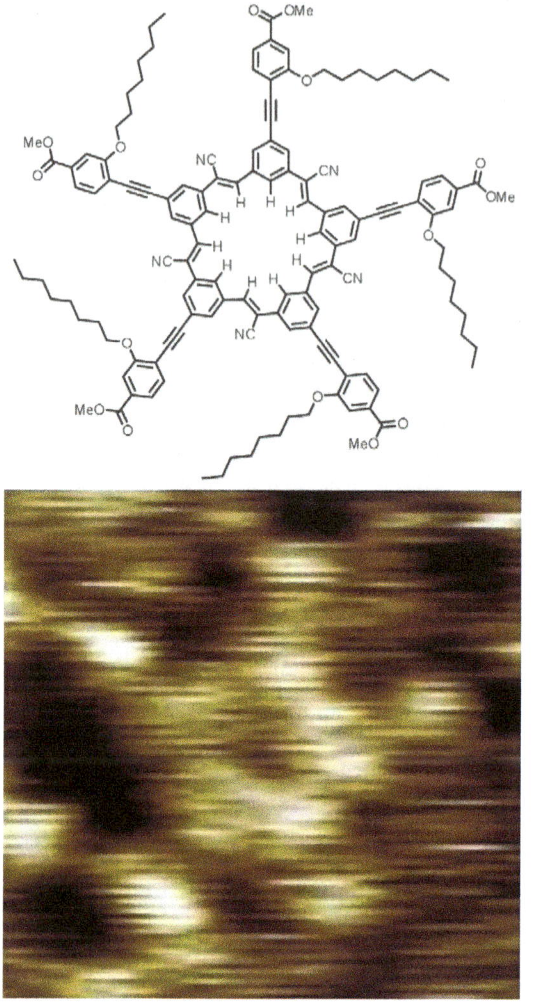

A first-generation "cyanostar" dendrimer and its STM image.

The first dendrimers were made by divergent synthesis approaches by Fritz Vögtle in 1978, R.G. Denkewalter at Allied Corporation in 1981, Donald Tomalia at Dow Chemical in 1983 and in 1985,and by George R. Newkome in 1985. In 1990 a convergent synthetic approach was introduced by Jean Fréchet. Dendrimer popularity then greatly increased, resulting in more than 5,000 scientific papers and patents by the year 2005.

Properties

Dendritic molecules are characterized by structural perfection. Dendrimers and dendrons are monodisperse and usually highly symmetric, spherical compounds. The field of dendritic molecules can be roughly divided into low-molecular weight and high-molecular weight species. The first category includes dendrimers and dendrons, and the latter includes dendronized polymers, hyperbranched polymers, and the polymer brush.

The properties of dendrimers are dominated by the functional groups on the molecular surface, however, there are examples of dendrimers with internal functionality. Dendritic encapsulation of functional molecules allows for the isolation of the active site, a structure that mimics that of active sites in biomaterials. Also, it is possible to make dendrimers water-soluble, unlike most polymers, by functionalizing their outer shell with charged species or other hydrophilic groups. Other controllable properties of dendrimers include toxicity, crystallinity, tecto-dendrimer formation, and chirality.

Dendrimers are also classified by generation, which refers to the number of repeated branching cycles that are performed during its synthesis. For example, if a dendrimer is made by convergent synthesis (see below), and the branching reactions are performed onto the core molecule three times, the resulting dendrimer is considered a third generation dendrimer. Each successive generation results in a dendrimer roughly twice the molecular weight of the previous generation. Higher generation dendrimers also have more exposed functional groups on the surface, which can later be used to customize the dendrimer for a given application.

Synthesis

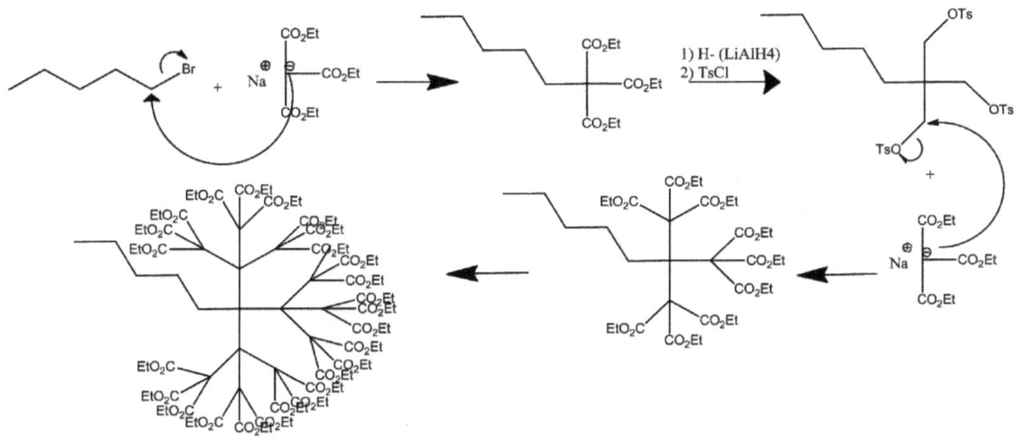

Synthesis to second generation arborol.

One of the very first dendrimers, the Newkome dendrimer, was synthesized in 1985. This macromolecule is also commonly known by the name arborol. The figure outlines the mechanism of the first two generations of arborol through a divergent route (discussed below). The synthesis is started by nucleophilic substitution of 1-bromopentane by triethyl sodiomethanetricarboxylate in dimethylformamide and benzene. The ester groups were then reduced by lithium aluminium hydride to a triol in a deprotection step. Activation of the chain ends was achieved by converting the alcohol groups to tosylate groups with tosyl chloride and pyridine. The tosyl group then served as leaving groups in another reaction with the tricarboxylate, forming generation two. Further repetition of the two steps leads to higher generations of arborol.

Poly(amidoamine), or PAMAM, is perhaps the most well known dendrimer. The core of PAMAM is a diamine (commonly ethylenediamine), which is reacted with methyl acrylate, and then another ethylenediamine to make the generation-0 (G-0) PAMAM. Successive reactions create higher generations, which tend to have different properties. Lower generations can be thought of as flexible molecules with no appreciable inner regions, while medium-sized (G-3 or G-4) do have internal space that is essentially separated from the outer shell of the dendrimer. Very large (G-7 and greater) dendrimers can be thought of more like solid particles with very dense surfaces due to the structure of their outer shell. The functional group on the surface of PAMAM dendrimers is ideal for click chemistry, which gives rise to many potential applications.

Dendrimers can be considered to have three major portions: a core, an inner shell, and an outer shell. Ideally, a dendrimer can be synthesized to have different functionality in each of these portions to control properties such as solubility, thermal stability, and attachment of compounds for particular applications. Synthetic processes can also precisely control the size and number of branches on the dendrimer. There are two defined methods of dendrimer synthesis, divergent synthesis and convergent synthesis. However, because the actual reactions consist of many steps needed to protect the active site, it is difficult to synthesize dendrimers using either method. This makes dendrimers hard to make and very expensive to purchase. At this time, there are only a few companies that sell dendrimers; Polymer Factory Sweden AB commercializes biocompatible bis-MPA dendrimers and Dendritech is the only kilogram-scale producers of PAMAM dendrimers. NanoSynthons, LLC from Mount Pleasant, Michigan, USA produces PAMAM dendrimers and other proprietary dendrimers.

Divergent Methods

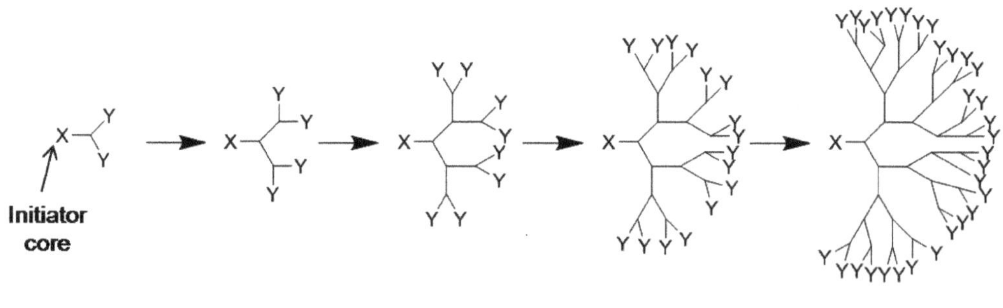

Schematic of divergent synthesis of dendrimers.

The dendrimer is assembled from a multifunctional core, which is extended outward by a series of reactions, commonly a Michael reaction. Each step of the reaction must be driven to full completion to prevent mistakes in the dendrimer, which can cause trailing generations (some branches are shorter than the others). Such impurities can impact the functionality and symmetry of the dendrimer, but are extremely difficult to purify out because the relative size difference between perfect and imperfect dendrimers is very small.

Convergent Methods

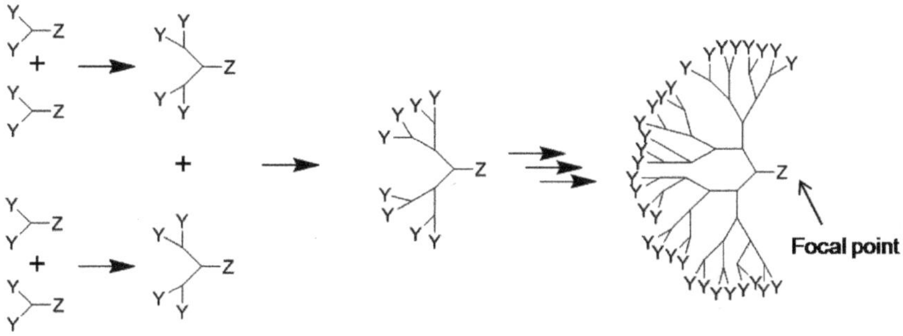

Schematic of convergent synthesis of dendrimers.

Dendrimers are built from small molecules that end up at the surface of the sphere, and reactions proceed inward building inward and are eventually attached to a core. This method makes it much easier to remove impurities and shorter branches along the way, so that the final dendrimer is more monodisperse. However dendrimers made this way are not as large as those made by divergent methods because crowding due to steric effects along the core is limiting.

Click Chemistry

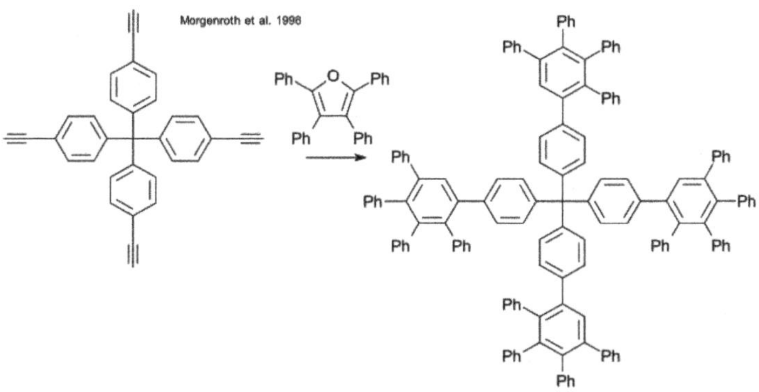

Dendrimer Diels-Alder reaction.

Dendrimers have been prepared via click chemistry, employing Diels-Alder reactions, thiol-ene and thiol-yne reactions and azide-alkyne reactions. There are ample avenues that can be opened by exploring this chemistry in dendrimer synthesis.

APPLICATIONS

Applications of dendrimers typically involve conjugating other chemical species to the dendrimer surface that can function as detecting agents (such as a dye molecule), affinity ligands, targeting components, radioligands, imaging agents, or pharmaceutically active compounds.

Dendrimers have very strong potential for these applications because their structure can lead to multivalent systems. In other words, one dendrimer molecule has hundreds of possible sites to couple to an active species. Researchers aimed to utilize the hydrophobic environments of the dendritic media to conduct photochemical reactions that generate the products that are synthetically challenged. Carboxylic acid and phenol-terminated water-soluble dendrimers were synthesized to establish their utility in drug delivery as well as conducting chemical reactions in their interiors. This might allow researchers to attach both targeting molecules and drug molecules to the same dendrimer, which could reduce negative side effects of medications on healthy cells. Dendrimers can also be used as a solubilizing agent. Since their introduction in the mid-1980s, this novel class of dendrimer architecture has been a prime candidate for host-guest chemistry. Dendrimers with hydrophobic core and hydrophilic periphery have shown to exhibit micelle-like behavior and have container properties in solution. The use of dendrimers as unimolecular micelles was proposed by Newkome in 1985. This analogy highlighted the utility of dendrimers as solubilizing agents. The majority of drugs available in pharmaceutical industry are hydrophobic in nature and this property in particular creates major formulation problems. This drawback of drugs can be ameliorated by dendrimeric scaffolding, which can be used to encapsulate as well as to solubilize the drugs because of the capability of such scaffolds to participate in extensive hydrogen bonding with water. Dendrimer labs throughout the planet are persistently trying to manipulate dendrimer's solubilizing trait, in their way to explore dendrimer as drug delivery and target specific carrier. For dendrimers to be able to be used in pharmaceutical applications, they must surmount the required regulatory hurdles to reach market. One dendrimer scaffold designed to achieve this is the Poly Ethoxy Ethyl Glycinamide (PEE-G) dendrimer. This dendrimer scaffold has been designed and shown to have high HPLC purity, stability, aqueous solubility and low inherent toxicity.

Drug Delivery

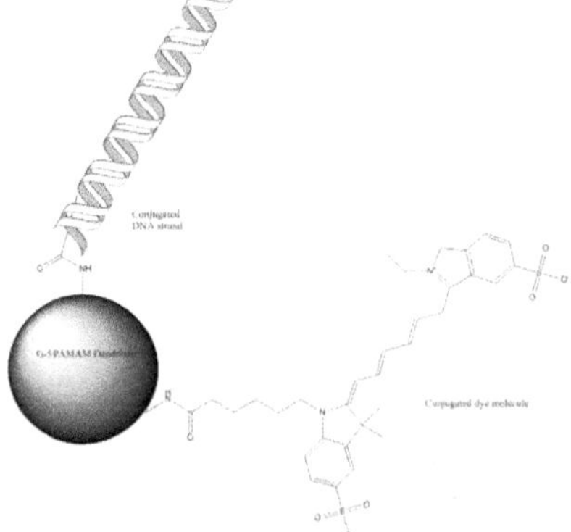

Schematic of a G-5 PAMAM dendrimer conjugated to both a dye molecule and a strand of DNA.

Approaches for delivering unaltered natural products using polymeric carriers is of widespread interest, dendrimers have been explored for the encapsulation of hydrophobic compounds and for the delivery of anticancer drugs. The physical characteristics of dendrimers, including their monodispersity, water solubility, encapsulation ability, and large number of functionalizable peripheral groups, make these macromolecules appropriate candidates for evaluation as drug delivery vehicles. There are three methods for using dendrimers in drug delivery: first, the drug is covalently attached to the periphery of the dendrimer to form dendrimer prodrugs, second the drug is coordinated to the outer functional groups via ionic interactions, or third the dendrimer acts as a unimolecular micelle by encapsulating a pharmaceutical through the formation of a dendrimer-drug supramolecular assembly. The use of dendrimers as drug carriers by encapsulating hydrophobic drugs is a potential method for delivering highly active pharmaceutical compounds that may not be in clinical use due to their limited water solubility and resulting suboptimal pharmacokinetics. Dendrimers have been widely explored for controlled delivery of antiretroviral bioactives The inherent antiretroviral activity of dendrimers enhances their efficacy as carriers for antiretroviral drugs. The dendrimer enhances both the uptake and retention of compounds within cancer cells, a finding that was not anticipated at the onset of studies. The encapsulation increases with dendrimer generation and this method may be useful to entrap drugs with a relatively high therapeutic dose. Studies based on this dendritic polymer also open up new avenues of research into the further development of drug-dendrimer complexes specific for a cancer and/or targeted organ system. These encouraging results provide further impetus to design, synthesize, and evaluate dendritic polymers for use in basic drug delivery studies and eventually in the clinic.

Gene Delivery

The ability to deliver pieces of DNA to the required parts of a cell includes many challenges. Current research is being performed to find ways to use dendrimers to traffic genes into cells without damaging or deactivating the DNA. To maintain the activity of DNA during dehydration, the dendrimer/DNA complexes were encapsulated in a water-soluble polymer, and then deposited on or sandwiched in functional polymer films with a fast degradation rate to mediate gene transfection. Based on this method, PAMAM dendrimer/DNA complexes were used to encapsulate functional biodegradable polymer films for substratemediated gene delivery. Research has shown that the fast-degrading functional polymer has great potential for localized transfection.

Sensors

Dendrimers have potential applications in sensors. Studied systems include proton or pH sensors using poly(propylene imine), cadmium-sulfide/polypropylenimine tetrahexacontaamine dendrimer composites to detect fluorescence signal quenching, and poly(propylenamine) first and second generation dendrimers for metal cation photodetection amongst others. Research in this field is vast and ongoing due to the potential for multiple detection and binding sites in dendritic structures.

Blood Substitution

Dendrimers are also being investigated for use as blood substitutes. Their steric bulk surrounding a heme-mimetic centre significantly slows degradation compared to free heme, and prevents the cytotoxicity exhibited by free heme.

Nanoparticles

Dendrimers also are used in the synthesis of monodisperse metallic nanoparticles. Poly(amidoamide), or PAMAM, dendrimers are utilized for their tertiary amine groups at the branching points within the dendrimer. Metal ions are introduced to an aqueous dendrimer solution and the metal ions form a complex with the lone pair of electrons present at the tertiary amines. After complexion, the ions are reduced to their zerovalent states to form a nanoparticle that is encapsulated within the dendrimer. These nanoparticles range in width from 1.5 to 10 nanometers and are called dendrimer-encapsulated nanoparticles.

Crop Protection and Agrochemicals

Given the widespread use of pesticides, herbicides and insecticides in modern farming, dendrimers are also being used by companies to help improve the delivery of agrochemicals to enable healthier plant growth and to help fight plant diseases.

MAGNETIC NANOPARTICLES

Magnetic nanoparticles are a class of nanoparticle that can be manipulated using magnetic fields. Such particles commonly consist of two components, a magnetic material, often iron, nickel and cobalt, and a chemical component that has functionality. While nanoparticles are smaller than 1 micrometer in diameter (typically 5–500 nanometers), the larger microbeads are 0.5–500 micrometer in diameter. Magnetic nanoparticle clusters that are composed of a number of individual magnetic nanoparticles are known as magnetic nanobeads with a diameter of 50–200 nanometers. Magnetic nanoparticle clusters are a basis for their further magnetic assembly into magnetic nanochains. The magnetic nanoparticles have been the focus of much research recently because they possess attractive properties which could see potential use in catalysis including nanomaterial-based catalysts, biomedicine and tissue specific targeting, magnetically tunable colloidal photonic crystals, microfluidics, magnetic resonance imaging, magnetic particle imaging, data storage, environmental remediation, nanofluids, and optical filters, defect sensor and cation sensors.

Properties

The physical and chemical properties of magnetic nanoparticles largely depend on the synthesis method and chemical structure. In most cases, the particles range from 1 to 100 nm in size and may display superparamagnetism.

Types of Magnetic Nanoparticles

Oxides: Ferrites

Ferrite nanoparticles or iron oxide nanoparticles (iron oxides in crystal structure of maghemite or magnetite) are the most explored magnetic nanoparticles up to date. Once the ferrite particles become smaller than 128 nm they become superparamagnetic which prevents self agglomeration since they exhibit their magnetic behavior only when an external magnetic field is applied. The magnetic moment of ferrite nanoparticles can be greatly increased by controlled clustering of a number of individual superparamagnetic nanoparticles into superparamagnetic nanoparticle clusters, namely magnetic nanobeads. With the external magnetic field switched off, the remanence falls back to zero. Just like non-magnetic oxide nanoparticles, the surface of ferrite nanoparticles is often modified by surfactants, silica, silicones or phosphoric acid derivatives to increase their stability in solution.

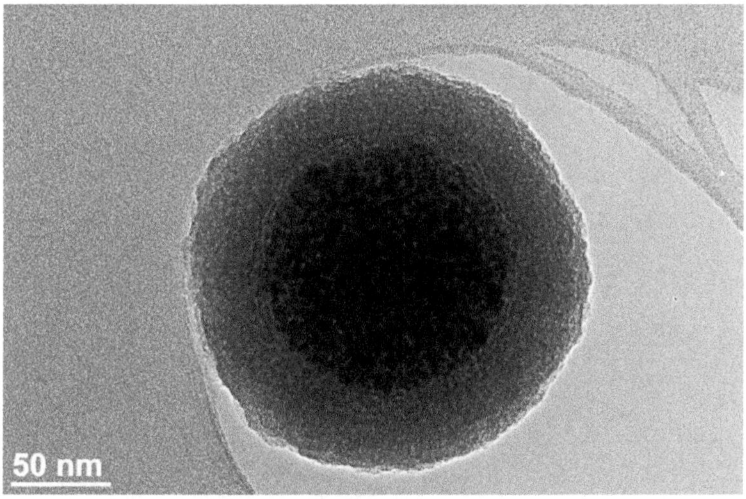

TEM image of a maghemite magnetic nanoparticle cluster with silica shell..

Ferrites with a shell

The surface of a maghemite or magnetite magnetic nanoparticle is relatively inert and does not usually allow strong covalent bonds with functionalization molecules. However, the reactivity of the magnetic nanoparticles can be improved by coating a layer of silica onto their surface. The silica shell can be easily modified with various surface functional groups via covalent bonds between organo-silane molecules and silica shell. In addition, some fluorescent dye molecules can be covalently bonded to the functionalized silica shell.

Ferrite nanoparticle clusters with narrow size distribution consisting of superparamagnetic oxide nanoparticles (~ 80 maghemite superparamagnetic nanoparticles per bead) coated with a silica shell have several advantages over metallic nanoparticles:

- Higher chemical stability (crucial for biomedical applications)

- Narrow size distribution (crucial for biomedical applications)
- Higher colloidal stability since they do not magnetically agglomerate
- Magnetic moment can be tuned with the nanoparticle cluster size
- Retained superparamagnetic properties (independent of the nanoparticle cluster size)
- Silica surface enables straightforward covalent functionalization

Metallic

Metallic nanoparticles may be beneficial for some technical applications due to their higher magnetic moment whereas oxides (maghemite, magnetite) would be beneficial for biomedical applications. This also implies that for the same moment, metallic nanoparticles can be made smaller than their oxide counterparts. On the other hand, metallic nanoparticles have the great disadvantage of being pyrophoric and reactive to oxidizing agents to various degrees. This makes their handling difficult and enables unwanted side reactions which makes them less appropriate for biomedical applications. Colloid formation for metallic particles is also much more challenging.

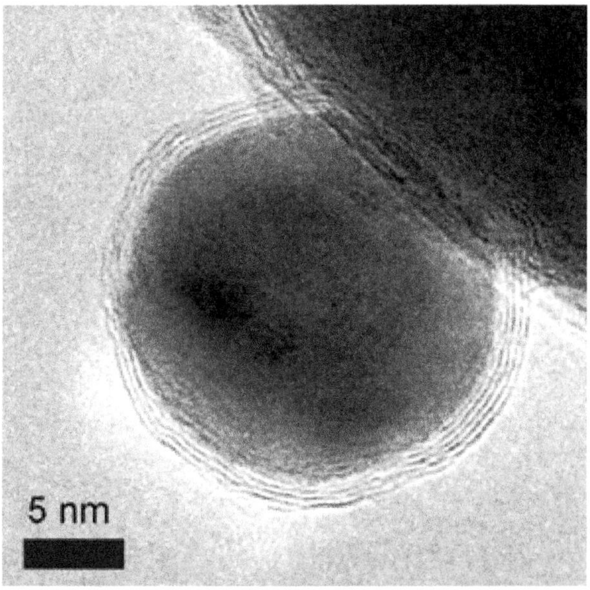

Cobalt nanoparticle with graphene shell (note: The individual graphene layers are visible)

Metallic with a shell

The metallic core of magnetic nanoparticles may be passivated by gentle oxidation, surfactants, polymers and precious metals. In an oxygen environment, Co nanoparticles form an anti-ferromagnetic CoO layer on the surface of the Co nanoparticle. Recently, work has explored the synthesis and exchange bias effect in these Co core CoO shell nanoparticles with a gold outer shell. Nanoparticles with a magnetic core consisting either of elementary Iron or Cobalt with a nonreactive shell made of graphene have been synthesized recently. The advantages compared to ferrite or elemental nanoparticles are:

- Higher magnetization
- Higher stability in acidic and basic solution as well as organic solvents
- Chemistry on the graphene surface via methods already known for carbon nanotubes

Synthesis

Several methods exist for preparing magnetic nanoparticle.

Co-Precipitation

Co-precipitation is a facile and convenient way to synthesize iron oxides (either Fe_3O_4 or γ-Fe_2O_3) from aqueous Fe^{2+}/Fe^{3+}salt solutions by the addition of a base under inert atmosphere at room temperature or at elevated temperature. The size, shape, and composition of the magnetic nanoparticles very much depends on the type of salts used (e.g.chlorides, sulfates, nitrates), the Fe^{2+}/Fe^{3+} ratio, the reaction temperature, the pH value and ionic strength of the media, and the mixing rate with the base solution used to provoke the precipitation. The co-precipitation approach has been used extensively to produce ferrite nanoparticles of controlled sizes and magnetic properties. A variety of experimental arrangements have been reported to facilitate continuous and large–scale co–precipitation of magnetic particles by rapid mixing.Recently, the growth rate of the magnetic nanoparticles was measured in real-time during the precipitation of magnetite nanoparticles by an integrated AC magnetic susceptometer within the mixing zone of the reactants.

Thermal Decomposition

Magnetic nanocrystals with smaller size can essentially be synthesized through the thermal decomposition of alkaline organometallic compounds in high-boiling organic solvents containing stabilizing surfactants.

Microemulsion

Using the microemulsion technique, metallic cobalt, cobalt/platinum alloys, and gold-coated cobalt/platinum nanoparticles have been synthesized in reverse micelles of cetyltrimethlyammonium bromide, using 1-butanol as the cosurfactant and octane as the oil phase.,

Flame spray Synthesis

Using flame spray pyrolysis and varying the reaction conditions, oxides, metal or carbon coated nanoparticles are produced at a rate of > 30 g/h .

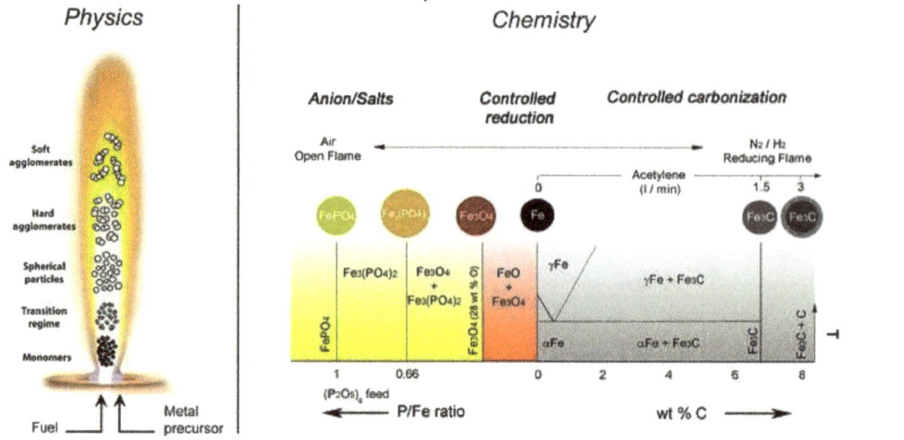

Operational layout differences between conventional and reducing flame spray synthesis

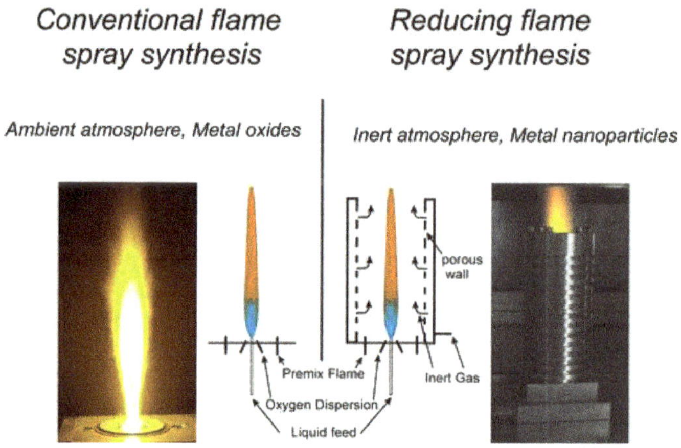

Various flame spray conditions and their impact on the resulting nanoparticles

Potential Applications

A wide variety of potential applications have been envisaged. Since magnetic nanoparticles are expensive to produce, there is interest in their recycling or for highly specialized applications. They are only used in scientific research. An industrial use has yet to be established.

The potential and versatility of magnetic chemistry arises from the fast and easy separation of the magnetic nanoparticles, eliminating tedious and costly separation processes usually applied in chemistry. Furthermore the magnetic nanoparticles can be guided via a magnetic field to the desired location which could, for example, enable pinpoint precision in fighting cancer.

Medical Diagnostics and Treatments

Magnetic nanoparticles have been examined for use in an experimental cancer treatment called magnetic hyperthermia in which an alternating magnetic field (AMF) is used to heat the nanoparticles. To achieve sufficient magnetic nanoparticle heating, the AMF typically has a frequency between 100-500kHz, although significant research has been done at lower frequencies as well as frequencies as high as 10MHz, with the amplitude of the field usually between 8-16kAm^{-1}.

Affinity ligands such as epidermal growth factor (EGF), folic acid, aptamers, lectins etc. can be attached to the magnetic nanoparticle surface with the use of various chemistries. This enables targeting of magnetic nanoparticles to specific tissues or cells. This strategy is used in cancer research to target and treat tumors in combination with magnetic hyperthermia or nanoparticle-delivered cancer drugs. Despite research efforts, however, the accumulation of nanoparticles inside of cancer tumors of all types is sub-optimal, even with affinity ligands. Willhelm et al. conducted a broad analysis of nanoparticle delivery to tumors and concluded that the median amount of injected dose reaching a solid tumor is only 0.7%. The challenge of accumulating large amounts of nanoparticles inside of tumors is arguably the biggest obstacle facing nanomedicine in general. While direct injection is used in some cases, intravenous injection is most often preferred to obtain a good distribution of particles throughout the tumor. Magnetic nanoparticles have a distinct advantage in that they can accumulate in desired regions via magnetically guided delivery, although this technique still needs further development to achieve optimal delivery to solid tumors. Another potential treatment of cancer includes attaching magnetic nanoparticles to free-floating cancer cells, allowing them to be captured and carried out of the body. The treatment has been tested in the laboratory on mice and will be looked at in survival studies. Magnetic nanoparticles can be used for the detection of cancer. Blood can be inserted onto a microfluidic chip with magnetic nanoparticles in it. These magnetic nanoparticles are trapped inside due to an externally applied magnetic field as the blood is free to flow through. The magnetic nanoparticles are coated with antibodies targeting cancer cells or proteins. The magnetic nanoparticles can be recovered and the attached cancer-associated molecules can be assayed to test for their existence.

Magnetic nanoparticles can be conjugated with carbohydrates and used for detection of bacteria. Iron oxide particles have been used for the detection of Gram negative bacteria like Escherichia coli and for detection of Gram positive bacteria like Streptococcus suis

Magnetic Immunoassay

Magnetic immunoassay (MIA) is a novel type of diagnostic immunoassay utilizing magnetic nanobeads as labels in lieu of conventional, enzymes, radioisotopes or fluorescent moieties. This assay involves the specific binding of an antibody to its antigen, where a magnetic label is conjugated to one element of the pair. The presence of magnetic nanobeads is then detected by a magnetic reader (magnetometer) which measures the magnetic field change induced by the beads. The signal measured by the magnetometer is proportional to the analyte (virus, toxin, bacteria, cardiac marker,etc.) quantity in the initial sample.

Waste water Treatment

Thanks to the easy separation by applying a magnetic field and the very large surface to volume ratio, magnetic nanoparticles have a potential for treatment of contaminated water. In this method, attachment of EDTA-like chelators to carbon coated metal nanomagnets results in a magnetic reagent for the rapid removal of heavy metals from solutions or contaminated water by three orders of magnitude to concentrations as low as micrograms per Litre. Magnetic nanobeads or nanoparticle clusters composed of FDA-approved oxide superparamagnetic nanoparticles (e.g. maghemite, magnetite) hold much potential for waste water treatment since they express excellent biocompatibility which concerning the environmental impacts of the material is an advantage compared to metallic nanoparticles.

Supported Enzymes and Peptides

Enzymes, proteins, and other biologically and chemically active substances have been immobilized on magnetic nanoparticles. They are of interest as possible supports for solid phase synthesis.

This technology is potentially relevant to cellular labelling/cell separation, detoxification of biological fluids, tissue repair, drug delivery, magnetic resonance imaging, hyperthermia and magnetofection.

Catalyst support

Magnetic nanoparticles are of potential use as a catalyst or catalyst supports. In chemistry, a catalyst support is the material, usually a solid with a high surface area, to which a catalyst is affixed. The reactivity of heterogeneous catalysts occurs at the surface atoms. Consequently, great effort is made to maximize the surface area of a catalyst by distributing it over the support. The support may be inert or participate in the catalytic reactions. Typical supports include various kinds of carbon, alumina, and silica. Immobilizing the catalytic center on top of nanoparticles with a large surface to volume ratio addresses this problem. In the case of magnetic nanoparticles it adds the property of facile a separation. An early example involved a rhodium catalysis attached to magnetic nanoparticles .

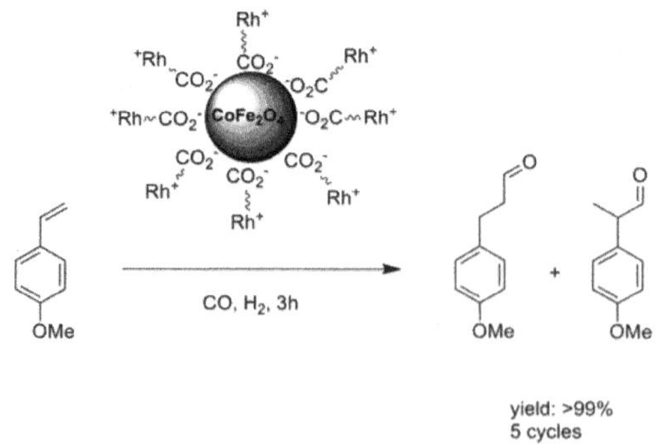

In another example, the stable radical TEMPO was attached to the graphene-coated cobalt nanoparticles via a diazonium reaction. The resulting catalyst was then used for the chemoselective oxidation of primary and secondary alcohols.

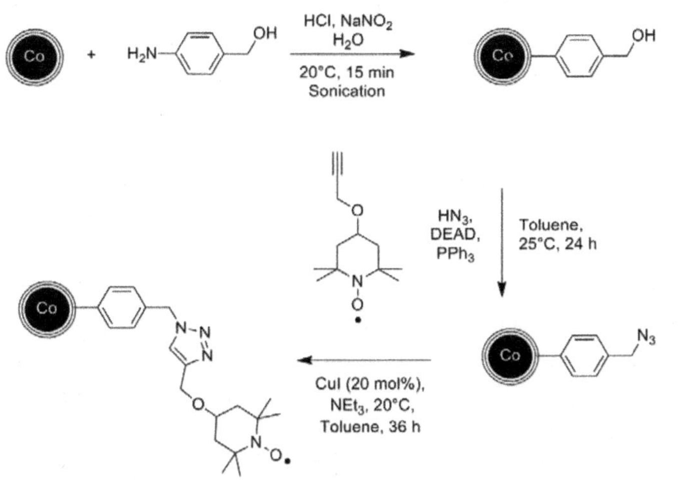

The catalytic reaction can be conducted in a continuous flow reactor instead of a batch reactor with no remains of the catalyst in the end product. Graphene coated cobalt nanoparticles have been used for that experiment since they exhibit a higher magnetization than Ferrite nanoparticles, which is essential for a fast and clean separation via external magnetic field.

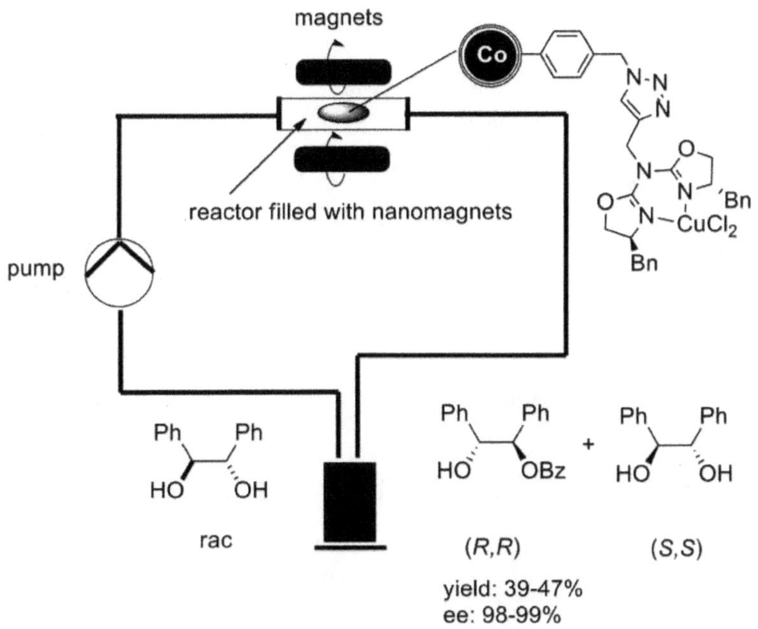

Biomedical Imaging

There are many applications for iron-oxide based nanoparticles in concert with magnetic resonance imaging. Magnetic CoPt nanoparticles are being used as an MRI contrast agent for transplanted neural stem cell detection.

Cancer Therapy

In magnetic fluid hyperthermia, nanoparticles of different types like Iron oxide, magnetite, maghemite or even gold are injected in tumor and then subjected under a high frequency magnetic field. These nanoparticles produce heat that typically increases tumor temperature to 40-46°C, which can kill cancer cells. Another major potential of magnetic nanoparticles is the ability to combine heat (hyperthermia) and drug release for a cancer treatment. Numerous studies have shown particle constructs that can be loaded with a drug cargo and magnetic nanoparticles. The most prevalent construct is the "Magnetoliposome", which is a liposome with magnetic nanoparticles typically embedded in the lipid bilayer. Under an alternating magnetic field, the magnetic nanoparticles are heated, and this heat permeabilizes the membrane. This causes release of the loaded drug. This treatment option has a lot of potential as the combination of hyperthermia and drug release is likely to treat tumors better than either option alone, but it is still under development.

Information Storage

A promising candidate for high-density storage is the face-centered tetragonal phase FePt alloy. Grain sizes can be as small as 3 nanometers. If it's possible to modify the MNPs at this small scale, the information density that can be achieved with this media could easily surpass 1 Terabyte per square inch.

Genetic Engineering

Magnetic nanoparticles can be used for a variety of genetics applications. One application is the rapid isolation of mRNA. In one application, the magnetic bead is attached to a poly T tail. When mixed with mRNA, the poly A tail of the mRNA will attach to the bead's poly T tail and the isolation takes place simply by placing a magnet on the side of the tube and pouring out the liquid. Magnetic beads have also been used in plasmid assembly. Rapid genetic circuit construction has been achieved by the sequential addition of genes onto a growing genetic chain, using nanobeads as an anchor. This method has been shown to be much faster than previous methods, taking less than an hour to create functional multi-gene constructs in vitro.

FUNGAL-DERIVED NANOPARTICLES

Throughout human history, fungi have been utilized as a source of food and harnessed to ferment and preserve foods and beverages. In the 20th century, humans have learned to harness fungi to protect human health (antibiotics, anti-cholesterol statins, and immunosuppressive agents), while industry has utilized fungi for large scale production of enzymes, acids, and biosurfactants. With the advent of modern nanotechnology in the 1980s, fungi have remained important by providing a greener alternative to chemically synthesized nanoparticle.

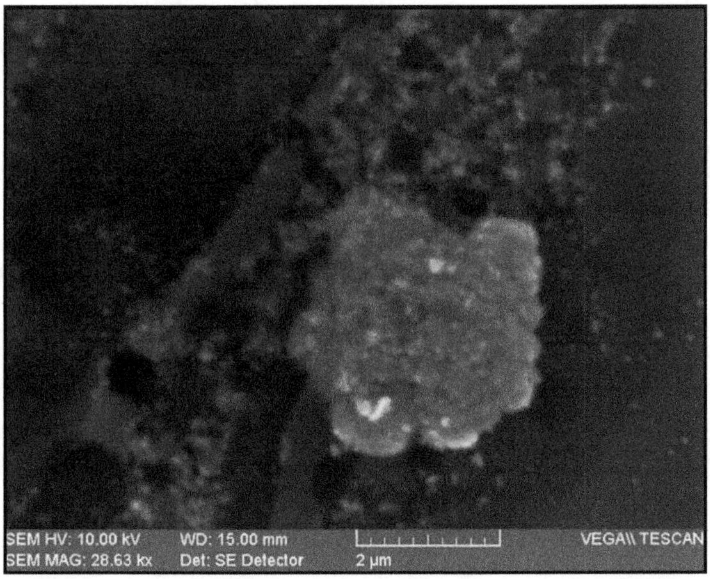

SEM image of fungal hyphae and fungal derived silver nanoparticles showing a large conglomeration made up of individual nanoparticles with fungal hyphae (dark areas) in background.

A nanoparticle is defined as having one dimension 100 nm or less in size. Environmentally toxic or biologically hazardous reducing agents are typically involved in the chemical synthesis of nanoparticles so there has been a search for greener production alternatives. Current research has shown that microorganisms, plant extracts, and fungi can produce nanoparticles through biological pathways. The most common nanoparticles synthesized by fungi are silver and gold, however fungi have been utilized in the synthesis other types of nanoparticles including zinc oxide, platinum, magnetite, zirconia, silica, titanium, and cadmium sulfide and cadmium selenide quantum dots.

Silver Nanoparticle Production

Synthesis of silver nanoparticles has been investigated utilizing many ubiquitous fungal species including Trichoderma, Fusarium,Penicillium, Rhizoctonia, Pleurotus and Aspergillus. Extracellular systhesis has been demonstrated by Trichoderma virde, T. reesei, Fusarium oxysporm, F. semitectum, F. solani, Aspergillus niger, A. flavus, A. fumigatus, A. clavatus, Pleurotus ostreatus, Cladosporium cladosporioides, Penicillium brevicompactum, P. fellutanum, an endophytic Rhizoctonia sp., Epicoccum nigrum, Chrysosporium tropicum, and Phoma glomerata, while intracellular synthesis was shown to occur in a Verticillium species, and in Neurospora crassa.

old Nanoparticle Production

Synthesis of gold nanoparticles has been investigated utilizing Fusarium, Neurospora, Verticillium, yeasts, and Aspergillus. Extracellular gold nanoparticle synthesis was demonstrated by Fusarium oxysporum, Aspergillus niger, and cytosolic extracts from Candida albican. Intracellular gold nanoparticle synthesis has been demonstrated by a Verticillum species, V. luteoalbum,

Miscellaneous Nanoparticle Production

In addition to gold and silver, Fusarium oxysporum has been used to synthesize zirconia, titanium, cadmium sulfide and cadmium selenide nanosize particles. Cadmium sulfide nanoparticles have also been synthesized by Trametes versicolor, Schizosaccharomyces pombe, and Candida glabrata. The whit-rot fungus Phanerochaete chrysosporium has also been demonstrated to be able to synthesize elemental selenium nanoparticles.

Culture Techniques and Conditions

Culture techniques and media vary depending upon the requirements of the fungal isolate involved, however the general procedure consist of the following: fungal hyphae are typically placed in liquid growth media and placed in shake culture until the fungal culture has increased in biomass. The fungal hyphae are removed from the growth media, washed with distilled water to remove the growth media, placed in distilled water and incubated on shake culture for 24 to 48 hours. The fungal hyphae are separated from the supernatant, and an aliquot of the supernatant is added to 1.0 mM ion solution. The ion solution is then monitored for 2 to 3 days for the formation of nanoparticles. Another common culture technique is to add washed fungal hyphae directly into 1.0 mM ion solution instead of utilizing the fungal filtrate. Silver nitrate is the most widely used source of silver ions, but silver sulfate has also been utilized. Choloroauric acid is generally used as the source of gold ions at various concentrations (1.0 mM and 250 mg to 500 mg of Au per liter). Cadmium sulfide nanoparticle synthesis for F. oxysporum was conducted using a 1:1 ratio of Cd2+ and SO42− at a 1 mM concentration. Gold nanoparticles can vary in shape and size depending on the pH of the ion solution. Gericke and Pinches (2006) reported that for V. luteoalbum small (cc.10 nm) spherical gold nanoparticles are formed at pH 3, larger (spherical, triangular, hexagon and rods) gold nanoparticles are formed at pH 5, and at pH 7 to pH 9 the large nanoparticles tend to lack a defined shape. Soni and Prakash (2011) reported an opposite response to pH for silver nanoparticles synthesized by both C. tropicum and F. oxysporum. The authors concluded that silver nanoparticle size increased as the pH of the ion solution decreased. Temperature interactions for both silver and gold nanoparticles were similar; a lower temperature resulted in larger nanoparticles while higher temperatures produced smaller nanoparticles.

Analytical Techniques

Visual Observations

For externally synthesized silver nanoparticles the silver ion solution generally becomes brownish in color, but this browning reaction may be absent. For fungi that synthesize intracellular silver nanoparticles, the hyphae darken to a brownish color while the solution remains clear. In both cases the browning reaction is attributed to the surface plasmon resonance of the metallic nanoparticles. For external gold nanoparticle production, the solution color can vary depending on the size of the gold nanoparticles; smaller particles appear pink while large particles appear purple. Intracellular gold nanoparticle synthesis typically turns the hyphae purple while the solution remains clear. Externally synthesized cadmium sulfide nanoparticles were reported to make the solution color appear bright yellow.

Analytical Tools

Scanning electron microscopy (SEM), transmission electron microscopy (TEM), energy dispersive analysis of X-ray (EDX), UV-vis spectroscopy, and X-ray diffraction are used to characterize different aspects of nanoparticles. Both SEM and TEM can be used to visualize the location, size, and morphology of the nanoparticles, while UV-vis spectroscopy can be used to confirm the metallic nature, size and aggregation level. Energy dispersive analysis of X-ray is used to determine elemental composition, and X-ray diffraction is used to determine chemical composition and crystallographic structure. UV-Vis absorption peaks for silver, gold, and cadmium sulfide nanoparticles can vary depending on particle size: 25-50 nm silver particles peak ca. 415 nm, gold nanoparticles 30-40 nm peak ca. 450 nm, while a cadmium sulfide absorption edge ca. 450 is indicative of quantum size particles. Larger nanoparticle of each type will have UV-Vis absorption peaks or edges that shift to longer wavelengths while smaller nanoparticles will have UV-Vis absorption peaks or edges that shift to shorter wavelengths.

Formation Mechanisms

Gold and silver

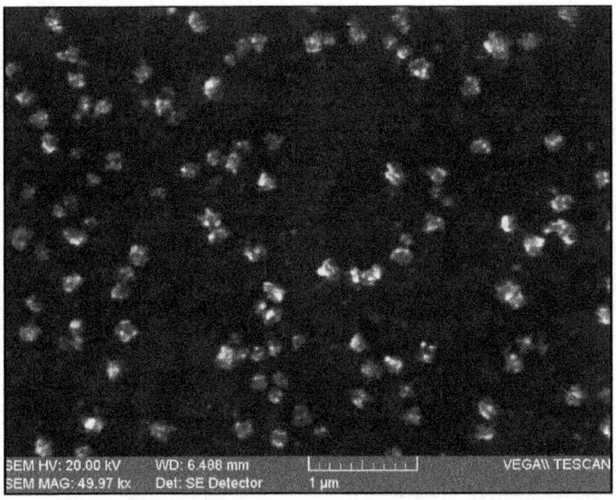

SEM image of fungal derived silver nanoparticles stabilized by a capping agent.

Nitrate reductase was suggested to initiate nanoparticle formation by many fungi including Penicillium species, while several enzymes, α-NADPH-dependent reductases, nitrate-dependent reductases and an extracellular shuttle quinone, were implicated in silver nanoparticle synthesis for Fusarium oxysporum. Jain et al. (2011) indicated that silver nanoparticle synthesis for A. flavus occurs initially by a "33kDa" protein followed by a protein (cystein and free amine groups) electrostatic attraction which stabilizes the nanoparticle by forming a capping agent. Intracellular silver and gold nanoparticle synthesis is not fully understood but similar fungal cell wall surface electrostatic attraction, reduction, and accumulation has been proposed. External gold nanoparticle synthesis by P. chrysosporium was attributed to laccase, while intracellular gold nanoparticle synthesis was attributed to ligninase.

Cadmium sulfide

Cadmium sulfide nanoparticle synthesis by yeast involves sequestration of Cd2+ by glutathione-related peptides followed by reduction within the cell. Ahmad et al. (2002) reported that cadmium sulfide nanoparticle synthesis by Fusarium oxysporum was based on a sulfate reductase (enzyme) process.

PLATINUM NANOPARTICLES

Platinum nanoparticles are usually in the form of a suspension or colloid of nanoparticles of platinum in a fluid, usually water. A colloid is technically defined as a stable dispersion of particles in a fluid medium (liquid or gas).

Spherical platinum nanoparticles can be made with sizes between about 2 and 100 nanometres (nm), depending on reaction conditions. Platinum nanoparticles are suspended in the colloidal solution of brownish-red or black color. Nanoparticles come in wide variety of shapes including spheres, rods, cubes, and tetrahedra.

Platinum nanoparticles are the subject of substantial research, with potential applications in a wide variety of areas. These include catalysis,medicine, and the synthesis of novel materials with unique properties.

Synthesis

Platinum nanoparticles are typically synthesized either by the reduction of platinum ion precursors in solution with a stabilizing or capping agent to form colloidal nanoparticles, or by the impregnation and reduction of platinum ion precursors in a micro-porous support such as alumina.

Some common examples of platinum precursors include potassium hexachloroplatinate (K2PtCl6) or platinous chloride (K2PtCl6) Different combinations of precursors, such as ruthenium chloride (RuCl3) and chloroplatinic acid (H2PtCl6), have been used to synthesize mixed-metal nanoparticles Some common examples of reducing agents include hydrogen gas (H2), sodium borohydride (NaBH4) and ethylene glycol (C2H6O2), although other alcohols and plant-derived compounds have also been used.

As the platinum metal precursor is reduced to neutral platinum metal (Pt0), the reaction mixture becomes supersaturated with platinum metal and the Pt0 begins to precipitate in the form of nanoscale particles. A capping agent or stabilizing agent such as sodium polyacrylic acid or sodium citrate is often used to stabilize the nanoparticle surfaces, and prevents the aggregation and coalescence of the nanoparticles.

The size of nanoparticles synthesized colloidally may be controlled by changing the platinum precursor, the ratio of capping agent to precursor, and/or the reaction temperature. The size of the nanoparticles can also be controlled with small deviation by using a stepwise seed-mediated growth procedure as outlined by Bigall et al. (2008). The size of nanoparticles synthesized onto a substrate such as alumina depends on various parameters such as the pore size of the support.

Platinum nanoparticles can also be synthesized by decomposing Pt2(dba)3 (dba = dibenzylideneacetone) under a CO or H2atmosphere, in the presence of a capping agent. The size and shape distributions of the resulting nanoparticles depend on the solvent, the reaction atmosphere, the types of capping agents and their relative concentrations, the specific platinum ion precursor, as well at the temperature of the system and reaction time.

Shape and Size Control

Ramirez et al. reported the influence of ligand and solvent effects on the size and shape of platinum nanoparticles. Platinum nanoparticle seeds were prepared by the decomposition of Pt2(dba)3 in tetrahydrofuran (THF) under carbon monoxide (CO). These conditions produced Pt nanoparticles with weakly bound THF and CO ligands and an approximate diameter on 1.2 nm. Hexadecylamine (HDA) was added to the purified reaction mixture and allowed to displace the THF and CO ligands over the course of approximately seven days, producing monodispersed spherical crystalline Pt nanoparticles with an average diameter of 2.1 nm. After the seven-day period, an elongation of the Pt nanoparticles occurred. When the same procedure was followed using a stronger capping agent such as triphenyl phosphine or octanethiol, the nanoparticles remained spherical, suggesting that the HDA ligand affects particle shape.

When Pt2(dba)3 was decomposed in THF under hydrogen gas in the presence HDA, the reaction took much longer, and formed nanowires with diameters between 1.5 and 2 nm. Decomposition of Pt2(dba)3 under hydrogen gas in toluene yielded the formation of nanowires with 2-3 nm diameter independent of HDA concentration. The length of these nanowires was found to be inversely proportional to the concentration of HDA present in solution. When these nanowire syntheses were repeated using reduced concentrations of Pt2(dba)3, there was little effect on the size, length or distribution of the nanowires formed. Platinum nanoparticles of controlled shape and size have also been accessed through varying the ratio of polymer capping agent concentration to precursor concentration. Reductive colloidal syntheses as such have yielded tetrahedral, cubic, irregular-prismatic, icosahedral, and cubo-octahedral nanoparticles, whose dispersity is also dependent on the concentration ratio of capping agent to precursor, and which may be applicable to catalysis. The precise mechanism of shape-controlled colloidal synthesis is not yet known; however, it is known that the relative growth rate of crystal facets within the growing nanostructure determines its final shape. Polyol syntheses of platinum nanoparticles, in which chloroplatinic acid is reduced to $PtCl_4^{2-}$ and Pt^0 by ethylene glycol, have also been a means to shape-controlled fabrication. Addition of varying amounts of sodium nitrate to these reactions was shown to yield tetrahedra and octahedra at high concentration ratios of sodium nitrate to chloroplatinic acid. Spectroscopic studies suggest that nitrate is reduced to nitrite by $PtCl_4^{2-}$ early in this reaction, and that the nitrite may then coordinate both Pt(II) and Pt(IV), greatly slowing the polyol reduction and altering the growth rates of distinct crystal facets within the nanoparticles, ultimately yielding morphological differentiation.

Green Synthesis

An ecologically-friendly synthesis of platinum nanoparticles from chloroplatinic acid was achieved through the use of a leaf extract of Diospyros kaki as a reducing agent. Nanoparticles synthesized as such were spherical with an average diameter ranging from 2-12 nm depending on reaction temperature and concentration of leaf extract used. Spectroscopic analysis suggests that this reaction is not enzyme-mediated and proceeds instead through plant-derived reductive small molecules.Another eco-friendly synthesis from chloroplatinic acid was reported using leaf extract from Ocimum sanctum and tulsi as reducing agents. Spectroscopic analysis suggested that ascorbic acid, gallic acid, various terpenes, and certain amino acids were active in the reduction. Particles synthesized as such were shown through scanning electron microscopy to consist in aggregates with irregular shape. It has been shown that tea extracts with high polyphenol content may be used both as reducing agents and capping agents for platinum nanoparticle synthesis.

Properties

The chemical and physical properties of platinum nanoparticles (NP) make them applicable for a wide variety of research applications. Extensive experimentation has been done to create new species of platinum NPs, and study their properties. Platinum NP applications include electronics, optics, catalysts, and enzyme immobilization.

Catalytic properties

Platinum NPs are used as catalysts for proton exchange membrane fuel cell (PEMFC), for industrial synthesis of nitric acid, reduction of exhaust gases from vehicles and as catalytic nucleating agents for synthesis of magnetic NPs. NPs can act as catalysts in homogeneous colloidal solution or as gas-phase catalysts while supported on solid state material. The catalytic reactivity of the NP is dependent on the shape, size and morphology of the particle

One type of platinum NPs that have been researched on are colloidal platinum NPs. Monometallic and bimetallic colloids have been used as catalysts in a wide range of organic chemistry, including, oxidation of carbon monoxide in aqueous solutions, hydrogenation of alkenes in organic or biphasic solutions and hydrosilylation of olefins in organic solutions. Collodial platinum NPs protected by Poly(N-isopropylacrylamide) were synthesised and their catalytic properties measured. It was determined that they were more active in solution and inactive when phase separated due to its solubility being inversely proportional to temperature.

Optical properties

Platinum NPs exhibit fascinating optical properties. Being a free electron metal NP like silver and gold, its linear optical response is mainly controlled by the surface plasmon resonance. Surface plasmon resonance occurs when the electrons in the metal surface are subject to an electromagnetic field that exerts a force on the electrons and cause them to displace from their original positions. The nuclei then exert a restoring force that results in oscillation of the electrons, which increase in strength when frequency of oscillations is in resonance with the incident electromagnetic wave.

The SPR of platinum nanoparticles is found in the ultraviolet range (215 nm), unlike the other noble metal nanoparticles which display SPR in the visible range. Experiments were done and the spectra obtained are similar for most platinum particles regardless of size. However, there is an exception. Platinum NPs synthesized via citrate reduction do not have a surface plasmon resonance peak around 215 nm. Through experimentation, the resonance peak only showed slight variations with the change of size and synthetic method (while maintaining the same shape), with the exception of those nanoparticles synthesized by citrate reduction, which did not exhibit and SPR peak in this region.

Through the control of percent composition of 2-5 nm platinum nanoparticles on SiO2, Zhang et al. modeled distinct absorption peaks attributed to platinum in the visible range, distinct from the conventional SPR absorption. This research attributed these absorption features to the generation and transfer of hot electrons from the platinum nanoparticles to the semiconductive material. Interestingly, the addition of small platinum nanoparticles on semiconductors such as TiO2 increases the photocatalytic oxidation activity under visible light irradiation. These concepts suggest the possible role of platinum nanoparticles in the development of solar energy conversion using metal nanoparticles. By changing the size, shape and environment of metal nanoparticles, their optical properties can be used for electrontic, catalytic, sensing, and photovoltaic applications.

Applications

Fuel cells application

Hydrogen Fuel Cells

Among the precious metals, platinum is the most active toward the hydrogen oxidation reaction that occurs at the anode in hydrogen fuel cells. In order to meet cost reductions of this magnitude, the Pt catalyst loading must be decreased. Two strategies have been investigated for reducing the Pt loading: the binary and ternary Pt-based alloyed nanomaterials and the dispersion of Pt-based nanomaterials onto high surface area substrates.

Methanol Fuel Cells

The methanol oxidation reaction occurs at the anode in direct methanol fuel cells (DMFCs). Platinum is the most promising candidate among pure metals for application in DMFCs. Platinum has the highest activity toward the dissociative adsorption of methanol. However, pure Pt surfaces are poisoned by carbon monoxide, a byproduct of methanol oxidation. Researchers have focused on dispersing nanostructured catalysts on high surface area supporting materials and the development of Pt-based nanomaterials with high electrocatalytic activity toward MOR to overcome the poisoning effect of CO.

Electrochemical Oxidation of Formic acid

Formic acid is another attractive fuel for use in PEM-based fuel cells. The dehydration pathway produces adsorbed carbon monoxide. A number of binary Pt-based nanomaterial electrocatalysts have been investigated for enhanced electrocatalytic activity toward formic acid oxidation.

Modifying Conductivity of Zinc Oxide Materials

Platinum NPs can be used to dope zinc oxide (ZnO) materials to improve their conductivity. ZnO has several characteristics that allow it to be used in several novel devices such as development of light-emitting assemblies and solar cells. However, because ZnO is of slightly lower conductivity than metal and indium tin oxide (ITO), it can be doped and hybridized with metal NPs like platinum to improve its conductivity. A method to do so would be to synthesize ZnO NPs using methanol reduction and incorporate at 0.25 at.% platinum NPs. This boosts the electrical properties of ZnO films while preserving its transmittance for application in transparent conducting oxides.

Glucose Detection Applications

Enzymatic glucose sensors have drawbacks that originate from the nature of the enzyme. Nonenzymatic glucose sensors with Pt-based electrocatalysts offer several advantages, including high stability and ease of fabrication. Many novel Pt and binary Pt-based nanomaterials have been developed to overcome the challenges of glucose oxidation on Pt surfaces, such as low selectivity, poor sensitivity, and poisoning from interfering species.

Other Applications

Platinum catalysts are alternatives of automotive catalytic converters, carbon monoxide gas sensors, petroleum refining, hydrogen production, and anticancer drugs. These applications utilize platinum nanomaterials due to their catalytic ability to oxidize CO and NOx, dehydrogenate hydrocarbons, and electrolyze water and their ability to inhibit the division of living cells.

Biological Interactions

The increased reactivity of nanoparticles is one of their most useful properties and is leveraged in fields such as catalysis, consumer products, and energy storage. However, this high reactivity also means that a nanoparticle in a biological environment may have unintended impacts. For example, many nanoparticles such as silver, copper, and ceria interact with cells to produce reactive oxygen species or ROS which can cause premature cell death through apoptosis. Determining the toxicity of a specific nanoparticle requires knowledge of the particle's chemical composition, shape, size and is a field that is growing alongside advances in nanoparticle research. Determining the impact of a nanoparticle on a living system is not straightforward. A multitude of in vivo and in vitro studies must be done to fully characterize reactivity. In vivo studies often use whole organisms such as mice or zebrafish to infer the interaction of the nanoparticle on a healthy human body. In vitro studies look at how nanoparticles interact with specific cell colonies, typically of human origin. Both types of experiments are needed for a complete understanding of nanoparticle toxicity, especially human toxicity, since no one model has complete human relevance.

Drug delivery

A topic of research within the field of nanoparticles is how to use these small particles for drug delivery. Depending on particle properties, nanoparticle may move throughout the human body are promising as site-specific vehicles for the transport of medicine. Current research using platinum nanoparticles in drug delivery uses platinum-based carries to move antitumor medicine. In one study, platinum nanoparticles of diameter 58.3 nm were used to transport an anticancer drug to human colon carcinoma cells, HT-29. Uptake of the nanoparticles by the cell involves compartmentalization of the nanoparticles within lysosomes. The high acidity environment enables leaching of platinum ions from the nanoparticle, which the researchers identified as causing the increased effectiveness of the drug. In another study, a Pt nanoparticle of diameter 140 nm was encapsulated within a PEG nanoparticle to move an antitumor drug, Cisplatin, within a prostate cancer cell (LNCaP/PC3) population. Use of platinum in drug delivery hinges on its ability to not interact in a harmful manner in healthy portions of the body while also being able to release its contents when in the correct environment.

Toxicology

Toxicity stemming from platinum nanoparticles can take multiple forms. One possible interaction is cytotoxicity or the ability of the nanoparticle to cause cell death. A nanoparticle can also interact with the cell's DNA or genome to cause genotoxicity.These effects are seen in different levels of gene expression measured through protein levels. Last is the developmental toxicity that can occur as an organism grows. Developmental toxicity looks at the impact the nanoparticle has on the growth of an organism from an embryonic stage to a later set point. Most nanotoxicology research is done on cyto- and genotoxicity as both can easily be done in a cell culture lab.

Platinum nanoparticles have the potential to be toxic to living cells. In one case, 2 nm platinum nanoparticles were exposed to two different types of algae in order to understand how these nanoparticles interact with a living system. In both species of algae tested, the platinum nanoparticles inhibited growth, induced small amounts of membrane damage, and created a large amount of oxidative stress. In another study, researcher tested the effects of differently sized platinum nanoparticles on primary human keratinocytes. The authors tested 5.8 and 57.0 nm Pt nanoparticles. The 57 nm nanoparticles had some hazardous effects including decreased cell metabolism, but the effect of the smaller nanoparticles was much more damaging. The 5.8 nm nanoparticles exhibited a more deleterious effect on the DNA stability of the primary keratincoytes than did the larger nanoparticles. The damage to the DNA was measured for individual cells using single-gel electrophoresis via the comet assay.

Researchers have also compared the toxicity of Pt nanoparticles to other commonly used metallic nanoparticles. In one study, the authors compared the impact of different nanoparticle compositions on the red blood cells found in the human bloodstream. The study showed that 5-10 nm platinum nanoparticles and 20-35 nm gold nanoparticles have very little effect on the red blood cells. In the same study it was found that 5-30 nm silver nanoparticles caused membrane damage, detrimental morphological variation, and haemagglutination to the red blood cells.

In a recent paper published in Nanotoxicology, the authors found that between silver (Ag-NP, d=5-35 nm), gold (Au-NP, d=15-35 nm), and Pt (Pt-NP, d=3-10 nm) nanoparticles, the Pt nanoparticles were the second most toxic in developing zebrafish embryos, behind only the Ag-NPs. However, this work did not examine the size dependence of the nanoparticles on their toxicity or biocompatibility. Size-dependent toxicity was determined by researchers at the National Sun Yat - Sen University in Kaohsiung, Taiwan. This group's work showed that the toxicity of platinum nanoparticles in bacterial cells is strongly dependent on nanoparticle size and shape/morphology. Their conclusions were based on two major observations. First, the authors found that platinum nanoparticles with spherical morphologies and sizes less than 3 nm showed biologically toxic properties; measured in terms of mortality, hatching delay, phenotypic defects and metal accumulation. While those nanoparticles with alternative shapes--such as cuboidal, oval, or floral--and sizes of 5-18 nm showed biocompatibility and no biologically toxic properties. Secondly, out of the three varieties of platinum nanoparticles which exhibited biocompatibility, two showed an increase in bacterial cell growth.

The paper introduces many hypotheses for why these observations were made, but based on other works and basic knowledge of bacterial cell membranes, the reasoning behind the size dependent toxicity observation seems to be two fold. One: The smaller, spherically shaped nanoparticles are able to pass through cell membranes simply due to their reduced size, as well as their shape-compatibility with the typically spherical pores of most cell membranes. Although this hypothesis needs to be further supported by future work, the authors did cite another paper which tracked the respiratory intake of platinum nanoparticles. This group found that 10 μm platinum nanoparticles are absorbed by the mucus of the bronchi and trachea, and can travel no further through the respiratory tract. However, 2.5 μm particles showed an ability to pass through this mucus layer, and reach much deeper into the respiratory tract. Also the larger, uniquely shaped nanoparticles are too large to pass through the pores of the cell membrane, and/or have shapes which are incompatible with the more spherically shaped pores of the cellular membrane. In regards to the observation that the two largest platinum nanoparticles (6-8 nm oval, and 16-18 nm floral) actually increase bacterial cell growth, the explanation could originate in the findings of other works which have shown that platinum nanoparticles have demonstrated significant antioxidative capacity. However, it must be noted that in order for these antioxidative properties to be exploited, the platinum nanoparticles must first enter the cells, so perhaps there is another explanation for this observation of increased bacterial cell growth.

Most studies so far have been size based using an in vivo mouse model. In one study, researchers compared the effects of sun 1 nm and 15 nm platinum nanoparticles on mice. The 15 mg/kg dose of sub 1nm platinum nanoparticles were found to cause liver damage while the larger particles had no effect. A similar study using a singular injection as an exposure source of platinum nanoparticles into the mouse model found necrosis of tubular epithelial cells for particles under 1 nm, but no effect with those particles of 8 nm. These in vivo studies show a trend that the toxicity of the platinum nanoparticles is size dependent, most likely due to the ability of the nanoparticle to get into a high impactful region within the body. A complete study analyzing the effect of varying sized platinum nanoparticles used both in vivo and in vitro models is used to gain a better understanding what impact these nanoparticles could have. Using mice as a model, they found retention of the platinum nanoparticles by the respiratory tract of the mouse. This was accompanied by a minor to mild inflammation of the surrounding lung tissue. However, their in vitro tests using human and lung epithelial cells found no cytotoxic or oxidative stress effects caused by the platinum nanoparticles despite clear evidence of cellular uptake.

NANOPARTICLE TRACKING ANALYSIS

Nanoparticle tracking analysis (NTA) is a method for visualizing and analyzing particles in liquids that relates the rate of Brownian motion to particle size. The rate of movement is related only to the viscosity and temperature of the liquid; it is not influenced by particle density or refractive index. NTA allows the determination of a size distribution profile of small particles with a diameter of approximately 10-1000 nanometers (nm) in liquid suspension.

The technique is used in conjunction with an ultramicroscope and a laser illumination unit that together allow small particles in liquid suspension to be visualized moving under Brownian motion. The light scattered by the particles is captured using a CCD or EMCCD camera over multiple frames. Computer software is then used to track the motion of each particle from frame to frame. The rate of particle movement is related to a sphere equivalent hydrodynamic radius as calculated through the Stokes–Einstein equation. The technique calculates particle size on a particle-by particle basis, overcoming inherent weaknesses in ensemble techniques such as dynamic light scattering. Since video clips form the basis of the analysis, accurate characterization of real time events such as aggregation and dissolution is possible. Samples require minimal preparation, minimizing the time required to process each sample. Speculators suggest that eventually the analysis may be done in real-time with no preparation, e.g. when detecting the presence of airborne viruses or biological weapons.

NTA currently operates for particles from about 10 to 1000 nm in diameter, depending on particle type. Analysis of particles at the lowest end of this range is possible only for particles composed of materials with a high refractive index, such gold and silver. The upper size limit is restricted by the limited Brownian motion of large particles; because a large particle moves very slowly, accuracy is diminished. The viscosity of the solvent also influences the movement of particles, and it, too, plays a part in determining the upper size limit for a specific system.

NTA was invented by Bob Carr, the founder and chief technology officer of NanoSight Ltd.

Applications

NTA has been used by commercial, academic, and government laboratories working with nanoparticle toxicology, drug delivery, exosomes, microvesicles, and other small biological particles, virology and vaccine production, ecotoxicology, protein aggregation, orthopedic implants, inks and pigments, and nanobubbles.

Comparison to Dynamic Light Scattering

Both dynamic light scattering (DLS) and nanoparticle tracking analysis (NTA) measure the Brownian motion of nanoparticles whose speed of motion, or diffusion constant, Dt, is related to particle size through the Stokes–Einstein equation.

$$\frac{(x,y)^2}{4} = Dt$$

$$Dt = \frac{TK_b}{3\pi\eta d} = Dt$$

where

- Dt is the diffusion constant, a product of diffusion coefficient D and time t
- K_b is Boltzmann's constant,
- T is the absolute temperature,
- η is viscosity
- d is the diameter of the spherical particle.

In NTA this motion is analyzed by video – individual particle positional changes are tracked in two dimensions from which the particle diffusion is determined. Knowing Dt, the particle hydrodynamic diameter can be then determined.

In contrast, DLS does not visualize the particles individually but analyzes, using a digital correlator, the time dependent scattering intensity fluctuations. These fluctuations are caused by interference effects arising from the relative Brownian movements of an ensemble of a large number of particles within a sample. Through analysis of the resultant exponential autocorrelation function, average particle size can be calculated as well as a polydispersity index. For multi-exponential autocorrelation functions arising from polydisperse samples, deconvolution can give limited information about the particle size distribution profile.

Iron oxide nanoparticles

Iron oxide nanoparticles are iron oxide particles with diameters between about 1 and 100 nanometers. The two main forms are magnetite (Fe3O4) and its oxidized form maghemite (γ-Fe2O3). They have attracted extensive interest due to their superparamagnetic properties and their potential applications in many fields (although Co and Ni are also highly magnetic materials, they are toxic and easily oxidized). Applications of iron oxide nanoparticles include terabit magnetic storage devices, catalysis, sensors, and high-sensitivity biomolecular magnetic resonance imaging (MRI) for medical diagnosis and therapeutics. These applications require coating of the nanoparticles by agents such as long-chain fatty acids, alkyl-substituted amines and diols.

Magnetic Properties

Due to its 4 unpaired electrons in 3d shell, an iron atom has a strong magnetic moment. Ions Fe2+ have also 4 unpaired electrons in 3d shell and Fe^{3+} have 5 unpaired electrons in 3d shell. Therefore, when crystals are formed from iron atoms or ions Fe^{2+} and Fe^{3+} they can be in ferromagnetic, antiferromagnetic or ferrimagnetic states. In the paramagnetic state, the individual atomic magnetic moments are randomly oriented, and the substance has a zero net magnetic moment if there is no magnetic field. These materials have a relative magnetic permeability greater than one and are attracted to magnetic fields. The magnetic moment drops to zero when the applied field is removed. But in a ferromagnetic material, all the atomic moments are aligned even without an external field. A ferrimagnetic material is similar to a ferromagnet but has two different types of atoms with opposing magnetic moments. The material has a magnetic moment because the opposing moments have different strengths. If they have the same magnitude, the crystal is antiferromagnetic and possesses no net magnetic moment. When an external magnetic field is applied to a ferromagnetic material, the magnetization (M) increases with the strength of the magnetic field (H) until it approaches saturation. Over some range of fields the magnetization has hysteresis because there is more than one stable magnetic state for each field. Therefore, a remanent magnetization will be present even after removing the external magnetic field. A single domain magnetic material (ex: magnetic nanoparticles) that has no hysteresis loop is said to be superparamagnetic. The ordering of magnetic moments in ferromagnetic, antiferromagnetic, and ferrimagnetic materials decreases with increasing temperature. Ferromagnetic and ferrimagnetic materials become disordered and lose their magnetization beyond the Curie temperature $T_C$and antiferromagnetic materials lose their magnetization beyond the Néel temperature T_N Magnetite is ferrimagnetic at room temperature and has a Curie temperature of 850 K. Maghemite is ferrimagnetic at room temperature, unstable at high temperatures, and loses its susceptibility with time. (Its Curie temperature is hard to determine). Both magnetite and maghemite nanoparticles are superparamagnetic at room temperature. This superparamagnetic behavior of iron oxide nanoparticles can be attributed to their size. When the size gets small enough (<10 nm), thermal fluctuations can change the direction of magnetization of the entire crystal. A material with many such crystals behaves like a paramagnet, except that the moments of entire crystals are fluctuating instead of individual atoms.

Synthesis

The preparation method has a large effect on shape, size distribution, and surface chemistry of the particles. It also determines to a great extent the distribution and type of structural defects or impurities in the particles. All these factors affect magnetic behavior. Recently, many attempts have been made to develop processes and techniques that would yield 'monodispersecolloids' consisting of nanoparticles uniform in size and shape.

Coprecipitation

By far the most employed method is coprecipitation. This method can be further divided into two types. In the first, ferrous hydroxide suspensions are partially oxidized with different oxidizing agents. For example, spherical magnetite particles of narrow size distribution with mean diameters between 30 and 100 nm can be obtained from a Fe(II) salt, a base and a mild oxidant (nitrate ions). The other method consists in ageing stoichiometric mixtures of ferrous and ferric hydroxides in aqueous media, yielding spherical magnetite particles homogeneous in size. In the second type, the following chemical reaction occurs:

$$2Fe^{3+} + Fe^{2+} + 8OH^{-} \longrightarrow Fe_3O_4 + 4H_2O$$

Optimum conditions for this reaction are pH between 8 and 14, Fe^{3+}/Fe^{2+} ratio of 2:1 and a non-oxidizing environment. Being highly susceptibile to oxidation, magnetite (Fe_3O_4) is transformed to maghemite (γFe_2O_3) in the presence of oxygen:

$$2Fe_3O_4 + O_2 \rightarrow 2\gamma Fe_2O_3$$

The size and shape of the nanoparticles can be controlled by adjusting pH, ionic strength, temperature, nature of the salts (perchlorates, chlorides, sulfates, and nitrates), or the Fe(II)/Fe(III) concentration ratio.

Microemulsions

A microemulsion is a stable isotropic dispersion of 2 immiscible liquids consisting of nanosized domains of one or both liquids in the other stabilized by an interfacial film of surface-active molecules. Microemulsions may be categorized further as oil-in-water (o/w) or water-in-oil (w/o), depending on the dispersed and continuous phases. Water-in-oil is more popular for synthesizing many kinds of nanoparticles. The water and oil are mixed with an amphiphillic surfactant. The surfactant lowers the surface tension between water and oil, making the solution transparent. The water nanodroplets act as nanoreactors for synthesizing nanoparticles. The shape of the water pool is spherical. The size of the nanoparticles will depend on size of the water pool to a great extent. Thus, the size of the spherical nanoparticles can be tailored and tuned by changing the size of the water pool.

High-Temperature Decomposition of Organic Precursors

The decomposition of iron precursors in the presence of hot organic surfactants results in samples with good size control, narrow size distribution (5-12 nm) and good crystallinity; and the nanoparticles are easily dispersed. For biomedical applications like magnetic resonance imaging, magnetic cell separation or magnetorelaxometry, where particle size plays a crucial role, magnetic nanoparticles produced by this method are very useful. Viable iron precursors include Fe(Cup)3,Fe(CO)5, or Fe(acac)3 in organic solvents with surfactant molecules. A combination of Xylenes and Sodium Dodecylbenezensulfonate as a surfactant are used to create nanoreactors for which well dispersed iron(II) and iron (III) salts can react.

Biomedical Applications

Magnetite and maghemite are preferred in biomedicine because they are biocompatible and potentially non-toxic to humans. Iron oxide is easily degradable and therefore useful for in vivo applications. Results from exposure of a human mesothelium cell line and a murine fibroblast cell line to seven industrially important nanoparticles showed a nanoparticle specific cytotoxic mechanism for uncoated iron oxide. Solubility was found to strongly influence the cytotoxic response. Labelling cells (e.g. stem cells, dendritic cells) with iron oxide nanoparticles is an interesting new tool to monitor such labelled cells in real time by magnetic resonance tomography.

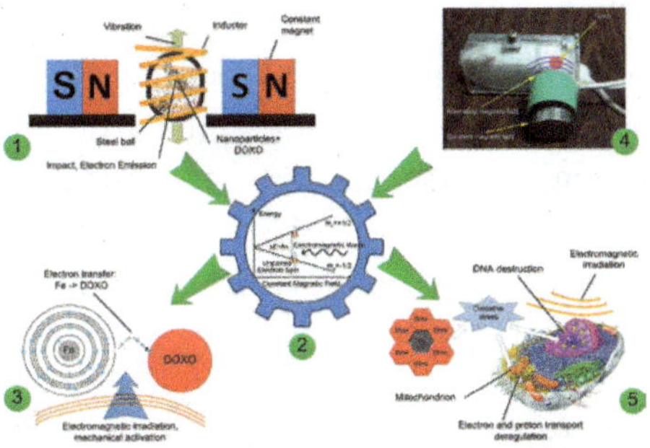

The magneto-mechano-chemical synthesis (1) is accompanied by splitting of electron energy levels (SEELs) and electron transfer in magnetic field (2) from nanoparticles Fe3O4 to doxorubicin. The concentration of paramagnetic centers (free radicals) is increased in the magneto-sensitive complex (MNC) (3). The local combined action of constant magnetic and electromagnetic fields and MNC in tumor (4) initiated SEELs, free radicals, leading to oxidative stress and electron and proton transport deregulation in the mitochondrion (5). Magnetic nanotherapy has more effectively inhibited the synthesis of ATP in mitochondria of tumor cell and induced the death of tumor cells compared to conventional doxorubicin.

Iron oxide nanoparticles are used in cancer magnetic nanotherapy that is based on the magneto-spin effects in free radical reactions and semiconductor material ability to generate oxygen radicals, furthermore, control oxidative stress in biological media under inhomogeneous electromagnetic radiation. The magnetic nanotherapy is remotely controlled by external electromagnetic field ROS (reactive oxygen species)-mediated local toxicity in the tumor during chemotherapy with antitumor magnetic complex and lesser side effects in normal tissues. Magnetic complexes that consist of iron oxide nanoparticles loaded with antitumor drug have additional advantages over conventional antitumor drugs due to their ability to be remotely controlled while targeting with a constant magnetic field and further strengthening of their antitumor activity by moderate inductive hyperthermia (below 40 °C).The combined influence of inhomogeneous constant magnetic and electromagnetic fields during nanotherapy has initiated splitting of electron energy levels in magnetic complex and unpaired electron transfer from iron oxide nanoparticles to anticancer drug and tumor cells. In particular, anthracycline antitumor antibiotic doxorubicin, the native state of which is diamagnetic, acquires the magnetic properties of paramagnetic substances. Electromagnetic radiation at the hyperfine splitting frequency can increase time radical pairs are in the triplet state and hence the probability of dissociation and so the concentration of free radicals. The reactivity of magnetic particles depends on their spin state. The experimental data was received about correlation between the frequency of electromagnetic field radiation with magnetic properties and quantity paramagnetic centres of complex. It is possible to control the kinetics of free radical reactions by external magnetic fields and modulate the level of oxidative stress (local toxicity) in malignant tumor. Cancer cells are then particularly vulnerable to an oxidative assault and induction of high levels of oxidative stress locally in tumor tissue, that has the potential to destroy or arrest the growth of cancer cells and can be thought as therapeutic strategy against cancer. Multifunctional magnetic complexes can combine cancer magnetic nanotherapy, tumor targeting and medical imaging functionalities in theranostics approach for personalized cancer medicine.

Iron oxide nanoparticles also are using in magnetic hyperthermia as a cancer treatment method. In this method, ferrofluid which is contained Iron oxide is injected to the tumor and then heated up by an alternating high frequency magnetic field. Temperature distribution produced by this heat generation destroys cancerous cells inside tumor.

REFERENCES

1. Module 3: Characteristics of Particles – Particle Size Categories. epa.gov
2. Taylor, Robert; Coulombe, Sylvain; Otanicar, Todd; Phelan, Patrick; Gunawan, Andrey; Lv, Wei; Rosengarten, Gary; Prasher, Ravi; Tyagi, Himanshu (2013). "Small particles, big impacts: A review of the diverse applications of nanofluids". Journal of Applied Physics. 113: 011301. Bibcode:2013JAP...113a1301T. doi:10.1063/1.4754271.
3. Taylor, Robert A; Otanicar, Todd; Rosengarten, Gary (2012). "Nanofluid-based optical filter optimization for PV/T systems". Light: Science & Applications. 1 (10): e34. doi:10.1038/lsa.2012.34.
4. Hewakuruppu, Y. L.; Dombrovsky, L. A.; Chen, C.; Timchenko, V.; Jiang, X.; Baek, S.; Taylor, R. A. (2013). "Plasmonic "pump–probe" method to study semi-transparent nanofluids". Applied Optics. 52 (24): 6041–6050. Bibcode:2013ApOpt..52.6041H. doi:10.1364/AO.52.006041. PMID 24085009.
5. , Robert A.; Otanicar, Todd P.; Herukerrupu, Yasitha; Bremond, Fabienne; Rosengarten, Gary; Hawkes, Evatt R.; Jiang, Xuchuan; Coulombe, Sylvain (2013). "Feasibility of nanofluid-based optical filters". Applied Optics. 52 (7): 1413–22. Bibcode:2013ApOpt..52.1413T. doi:10.1364/AO.52.001413. PMID 23458793.
6. Vert, M.; Doi, Y.; Hellwich, K. H.; Hess, M.; Hodge, P.; Kubisa, P.; Rinaudo, M.; Schué, F. O. (2012). "Terminology for biorelated polymers and applications (IUPAC Recommendations 2012)". Pure and Applied Chemistry. 84 (2). doi:10.1351/PAC-REC-10-12-04.

7. MacNaught, Alan D.; Wilkinson, Andrew R., eds. (1997). Compendium of Chemical Terminology: IUPAC Recommendations (2nd ed.). Blackwell Science. ISBN 0865426848.

8. Alemán, J.; Chadwick, A. V.; He, J.; Hess, M.; Horie, K.; Jones, R. G.; Kratochvíl, P.; Meisel, I.; Mita, I.; Moad, G.; Penczek, S.; Stepto, R. F. T. (2007). "Definitions of terms relating to the structure and processing of sols, gels, networks, and inorganic-organic hybrid materials (IUPAC Recommendations 2007)". Pure and Applied Chemistry. 79 (10): 1801. doi:10.1351/pac200779101801.

9. Vert, Michel; Doi, Yoshiharu; Hellwich, Karl-Heinz; Hess, Michael; Hodge, Philip; Kubisa, Przemyslaw; Rinaudo, Marguerite; Schué, François (2012). "Terminology for biorelated polymers and applications (IUPAC Recommendations 2012)" (PDF). Pure and Applied Chemistry. 84 (2): 377–410. doi:10.1351/PAC-REC-10-12-04.

10. Granqvist, C.; Buhrman, R.; Wyns, J.; Sievers, A. (1976). "Far-Infrared Absorption in Ultrafine Al Particles". Physical Review Letters. 37 (10): 625–629. Bibcode:1976PhRvL..37..625G. doi:10.1103/PhysRevLett.37.625.

11. Hayashi, C.; Uyeda, R & Tasaki, A. (1997). Ultra-fine particles: exploratory science and technology (1997 Translation of the Japan report of the related ERATO Project 1981–86). Noyes Publications.

12. Kiss, L. B.; Söderlund, J.; Niklasson, G. A.; Granqvist, C. G. (1999). "New approach to the origin of lognormal size distributions of nanoparticles". Nanotechnology. 10: 25–28. Bibcode:1999Nanot..10...25K. doi:10.1088/0957-4484/10/1/006.

13. Buzea, C.; Pacheco, I. I.; Robbie, K. (2007). "Nanomaterials and nanoparticles: Sources and toxicity". Biointerphases. 2 (4): MR17–MR71. doi:10.1116/1.2815690. PMID 20419892.

14. ASTM E 2456 – 06 Standard Terminology Relating to Nanotechnology

15. Fahlman, B. D. (2007). Materials Chemistry. Springer. pp. 282–283. ISBN 1-4020-6119-6. Retrieved 2016-12-06.

16. «Nanotechnology Timeline | Nano". www.nano.gov. Retrieved 2016-12-12.

17. Reiss, Gunter; Hutten, Andreas (2010). "Magnetic Nanoparticles". In Sattler, Klaus D. Handbook of Nanophysics: Nanoparticles and Quantum Dots. CRC Press. pp. 2–1. ISBN 9781420075458. Retrieved 2016-12-06.

18. Khan, Firdos Alam (2012). Biotechnology Fundamentals. CRC Press. p. 328. ISBN 9781439820094. Retrieved 2016-12-06.

19. Rawson, Philip S. (1984). Ceramics. University of Pennsylvania Press. ISBN 0-8122-1156-1. Retrieved 2016-12-06.

20. Faraday, Michael (1857). "Experimental relations of gold (and other metals) to light". Phil. Trans. Roy. Soc. London. 147: 145–181. doi:10.1098/rstl.1857.0011.

21. Beilby, G.T. (1903). "The Effects of Heat and of Solvents on Thin Films of Metal". Proceedings of the Royal Society A. 72 (477–486): 226–235. doi:10.1098/rspl.1903.0046. JSTOR 116470.

22. Turner, T. (1908). "Transparent Silver and Other Metallic Films". Proceedings of the Royal Society A (Transparent silver and other metallic films). 81 (548): 301–310. Bibcode:1908RSPSA..81..301T. doi:10.1098/rspa.1908.0084. JSTOR 93060.

23. Onoda, G.Y. Jr.; Hench, L.L., eds. (1979). Ceramic Processing Before Firing. New York: Wiley & Sons. ISBN 0-471-65410-8.

24. Aksay, I.A., Lange, F.F., Davis, B.I. (1983). "Uniformity of Al2O3-ZrO2 Composites by Colloidal Filtration". J. Am. Ceram. Soc. 66 (10): C–190. doi:10.1111/j.1151-2916.1983.tb10550.x.

25. Franks, G.V. & Lange, F.F. (1996). "Plastic-to-Brittle Transition of Saturated, Alumina Powder Compacts". J. Am. Ceram. Soc. 79 (12): 3161–3168. doi:10.1111/j.1151-2916.1996.tb08091.x.

26. Evans, A.G. & Davidge, R.W. (1969). "The strength and fracture of fully dense polycrystalline magnesium oxide". Phil. Mag. 20 (164): 373–388. Bibcode:1969PMag...20..373E. doi:10.1080/14786436908228708.

27. Evans, A. G.; Davidge, R. W. (1970). "The strength and oxidation of reaction-sintered silicon nitride". J Mat. Sci. 5 (4): 314–325. Bibcode:1970JMatS...5..314E. doi:10.1007/BF02397783.

28. Lange, F.F. & Metcalf, M. (1983). "Processing-Related Fracture Origins: II, Agglomerate Motion and Cracklike Internal Surfaces Caused by Differential Sintering". J. Am. Ceram. Soc. 66 (6): 398–406. doi:10.1111/j.1151-2916.1983.tb10069.x.

29. Evans, A.G. (1987). "Considerations of Inhomogeneity Effects in Sintering". J. Am. Ceram. Soc. 65 (10): 497–501. doi:10.1111/j.1151-2916.1982.tb10340.x.

30. Whitesides, G.M.; et al. (1991). "Molecular Self-Assembly and Nanochemistry: A Chemical Strategy for the Synthesis of Nanostructures". Science. 254 (5036): 1312–9. Bibcode:1991Sci...254.1312W. doi:10.1126/science.1962191. PMID 1962191.

31. Dabbs D. M, Aksay I.A.; Aksay (2000). "Self-Assembled Ceramics". Annu. Rev. Phys. Chem. 51: 601–22. Bibcode:2000ARPC...51..601D. doi:10.1146/annurev.physchem.51.1.601. PMID 11031294.

32. Buffat, Ph.; Borel, J.-P. (1976). "Size effect on the melting temperature of gold particles". Physical Review A. 13 (6): 2287–2298. Bibcode:1976PhRvA..13.2287B. doi:10.1103/PhysRevA.13.2287.

33. Taylor, Robert A; Phelan, Patrick E; Otanicar, Todd P; Adrian, Ronald; Prasher, Ravi (2011). "Nanofluid optical property characterization: Towards efficient direct absorption solar collectors". Nanoscale Research Letters. 6 (1): 225. Bibcode:2011NRL.....6..225T. doi:10.1186/1556-276X-6-225. PMC 3211283 . PMID 21711750.

34. Gubin, Sergey P. (2009). Magnetic nanoparticles. Wiley-VCH. ISBN 3-527-40790-1.

35. «Sunscreen". U.S. Food and Drug Administration. Retrieved 2016-12-06.

36. Mitchnick, MA; Fairhurst, D; Pinnell, SR (1999). "Microfine zinc oxide (Z-cote) as a photostable UVA/UVB sunblock agent". Journal of the American Academy of Dermatology. 40 (1): 85–90. doi:10.1016/S0190-9622(99)70532-3. PMID 9922017.

37. Heim, J; Felder, E; Tahir, MN; Kaltbeitzel, A; Heinrich, UR; Brochhausen, C; Mailänder, V; Tremel, W; Brieger, J (21 May 2015). "Genotoxic effects of zinc oxide nanoparticles.". Nanoscale. 7 (19): 8931–8. Bibcode:2015Nanos...7.8931H. doi:10.1039/c5nr01167a. PMID 25916659.

38. Wang, B; Zhang, Y; Mao, Z; Yu, D; Gao, C (August 2014). "Toxicity of ZnO nanoparticles to macrophages due to cell uptake and intracellular release of zinc ions.". Journal of Nanoscience and Nanotechnology. 14 (8): 5688–96. doi:10.1166/jnn.2014.8876. PMID 25935990.

39. Gosens, I; Kermanizadeh, A; Jacobsen, NR; Lenz, AG; Bokkers, B; de Jong, WH; Krystek, P; Tran, L; Stone, V; Wallin, H; Stoeger, T; Cassee, FR (2015). "Comparative hazard identification by a single dose lung exposure of zinc oxide and silver nanomaterials in mice.". PLOS ONE. 10 (5): e0126934. doi:10.1371/journal.pone.0126934. PMC 4429007. PMID 25966284.

40. Hanagata, N; Morita, H (2015). "Calcium ions rescue human lung epithelial cells from the toxicity of zinc oxide nanoparticles.". The Journal of toxicological sciences. 40 (5): 625–35. doi:10.2131/jts.40.625. PMID 26354379.

41. Kim, YH; Kwak, KA; Kim, TS; Seok, JH; Roh, HS; Lee, JK; Jeong, J; Meang, EH; Hong, JS; Lee, YS; Kang, JS (June 2015). "Retinopathy Induced by Zinc Oxide Nanoparticles in Rats Assessed by Micro-computed Tomography and Histopathology.". Toxicological research. 31 (2): 157–63. doi:10.5487/tr.2015.31.2.157. PMID 26191382.

42. Moridian, M; Khorsandi, L; Talebi, AR (2015). "Morphometric and stereological assessment of the effects of zinc oxide nanoparticles on the mouse testicular tissue.". Bratislavske lekarske listy. 116 (5): 321–5. doi:10.4149/bll_2015_060. PMID 25924642.

43. «The Textiles Nanotechnology Laboratory". nanotextiles.human.cornell.edu. Retrieved 2016-12-06.

44. Belloni, J.; Mostafavi, M.; Remita, H.; Marignier, J. L.; Delcourt, A. M. O. (1998). "Radiation-induced synthesis of mono- and multi-metallic clusters and nanocolloids". New Journal of Chemistry. 22 (11): 1239–1255. doi:10.1039/A801445K.

45. Brinker, C.J. & Scherer, G.W. (1990). Sol-Gel Science: The Physics and Chemistry of Sol-Gel Processing. Academic Press. ISBN 0-12-134970-5.

46. Hench, L. L.; West, J. K. (1990). "The sol-gel process". Chemical Reviews. 90: 33–72. doi:10.1021/cr00099a003.

47. Klein, L. (1994). Sol-Gel Optics: Processing and Applications. Springer Verlag. ISBN 0-7923-9424-0. Retrieved 2016-12-06.

48. Corriu, Robert & Anh, Nguyên Trong (2009). Molecular Chemistry of Sol-Gel Derived Nanomaterials. John Wiley and Sons. ISBN 0-470-72117-0. Retrieved

2016-12-06.

49. Pais, A. (2005). Subtle is the Lord: The Science and the Life of Albert Einstein. Oxford University Press. ISBN 0-19-280672-6. Retrieved 2016-12-06.

50. Agam, M. A.; Guo, Q (2007). "Electron Beam Modification of Polymer Nanospheres". Journal of Nanoscience and Nanotechnology. 7 (10): 3615–9. doi:10.1166/jnn.2007.814. PMID 18330181.

51. Kralj, Slavko; Makovec, Darko (27 October 2015). "Magnetic Assembly of Superparamagnetic Iron Oxide Nanoparticle Clusters into Nanochains and Nanobundles". ACS Nano. 9 (10): 9700–9707. doi:10.1021/acsnano.5b02328.

52. Choy J.H.; Jang E.S.; Won J.H.; Chung J.H.; Jang D.J. & Kim Y.W. (2004). "Hydrothermal route to ZnO nanocoral reefs and nanofibers". Appl. Phys. Lett. 84 (2): 287. Bibcode:2004ApPhL..84..287C. doi:10.1063/1.1639514.

53. Sun, Y; Xia, Y; Xia (2002). "Shape-controlled synthesis of gold and silver nanoparticles". Science. 298 (5601): 2176–9. Bibcode:2002Sci...298.2176S. doi:10.1126/science.1077229. PMID 12481134.

54. Murphy, C.J. (2002). "Materials science. Nanocubes and nanoboxes". Science. 298 (5601): 2139–41. doi:10.1126/science.1080007. PMID 12481122.

55. Anderson, Will; Kozak, Darby; Coleman, Victoria; Jamting, Asa; Trau, Matt (1 September 2013), "A comparative study of submicron particle sizing platforms: Accuracy, precision and resolution analysis of polydisperse particle size distributions", Journal of Colloid and Interface Science, UK, 405: 322–330, doi:10.1016/j.jcis.2013.02.030

56. Prime, KL; Whitesides, GM (1991). "Self-assembled organic monolayers: model systems for studying adsorption of proteins at surfaces". Science. 252 (5009): 1164–7. Bibcode:1991Sci...252.1164P. doi:10.1126/science.252.5009.1164. PMID 2031186.

57. Liu, W; Greytak, AB; Lee, J; Wong, CR; Park, J; Marshall, LF; Jiang, W; Curtin, PN; Ting, AY; Nocera, Daniel G.; Fukumura, Dai; Jain, Rakesh K.; Bawendi, Moungi G. (2010). "Compact biocompatible quantum dots via RAFT-mediated synthesis of imidazole-based random copolymer ligand". Journal of the American Chemical Society. 132 (2): 472–83. doi:10.1021/ja908137d. PMC 2871316. PMID 20025223.

58. Akerman ME, Chan WC, Laakkonen P, Bhatia SN, Ruoslahti E.; Chan; Laakkonen; Bhatia; Ruoslahti (2002). "Nanocrystal targeting in vivo". Proceedings of the National Academy of Sciences of the United States of America. 99 (20): 12617–21. Bibcode:2002PNAS...9912617A. doi:10.1073/pnas.152463399. PMC 130509. PMID 12235356.

59. Hoshino, A; Fujioka, K; Oku, T; Nakamura, S; Suga, M; Yamaguchi, Y; Suzuki, K; Yasuhara, M; Yamamoto, K (2004). "Quantum dots targeted to the assigned organelle in living cells". Microbiology and immunology. 48 (12): 985–94. doi:10.1111/j.1348-0421.2004.tb03621.x. PMID 15611617.

60. Suzuki, KG; Fujiwara, TK; Edidin, M; Kusumi, A (2007). "Dynamic recruitment of phospholipase Cγ at transiently immobilized GPI-anchored receptor clusters

induces IP3–Ca2+ signaling: single-molecule tracking study 2". The Journal of Cell Biology. 177 (4): 731–42. doi:10.1083/jcb.200609175. PMC 2064217. PMID 17517965.

61. Sung, KM; Mosley, DW; Peelle, BR; Zhang, S; Jacobson, JM (2004). "Synthesis of monofunctionalized gold nanoparticles by fmoc solid-phase reactions". Journal of the American Chemical Society. 126 (16): 5064–5. doi:10.1021/ja049578p. PMID 15099078.

62. Fu, A; Micheel, CM; Cha, J; Chang, H; Yang, H; Alivisatos, AP (2004). "Discrete nanostructures of quantum dots/Au with DNA". Journal of the American Chemical Society. 126 (35): 10832–3. doi:10.1021/ja046747x. PMID 15339154.

63. Howarth, M; Liu, W; Puthenveetil, S; Zheng, Y; Marshall, LF; Schmidt, MM; Wittrup, KD; Bawendi, MG; Ting, AY (2008). "Monovalent, reduced-size quantum dots for imaging receptors on living cells". Nature Methods. 5 (5): 397–9. doi:10.1038/nmeth.1206. PMC 2637151. PMID 18425138.

64. Nanoparticles play at being red blood cells

65. Mnyusiwalla, Anisa; Daar, Abdallah S; Singer, Peter A (2003). "Mind the gap : science and ethics in nanotechnology". Nanotechnology. 14 (3): R9. Bibcode:2003Nanot..14R...9M. doi:10.1088/0957-4484/14/3/201.

66. «Toxic Nanoparticles Might be Entering Human Food Supply, MU Study Finds". University of Missouri. 22 August 2013. Retrieved 23 August 2013.

67. Ying, Jackie (2001). Nanostructured Materials. New York: Academic Press. ISBN 0-12-744451-3. Retrieved 2016-12-06.

68. Nanotechnologies: 6. What are potential harmful effects of nanoparticles? europa.eu

69. Greulich, C.; Diendorf, J.; Simon, T.; Eggeler, G.; Epple, M.; Köller, M. (2011). "Uptake and intracellular distribution of silver nanoparticles in human mesenchymal stem cells". Acta Biomaterialia. 7 (1): 347–354. doi:10.1016/j.actbio.2010.08.003. ISSN 1742-7061.

70. Hanley, C; Thurber, A; Hanna, C; Punnoose, A; Zhang, J; Wingett, DG; Thurber; Hanna; Punnoose; Zhang; Wingett (2009). "The Influences of Cell Type and ZnO Nanoparticle Size on Immune Cell Cytotoxicity and Cytokine Induction". Nanoscale Res Lett. 4 (12): 1409–20. Bibcode:2009NRL.....4.1409H. doi:10.1007/s11671-009-9413-8. PMC 2894345. PMID 20652105.

71. Vines T, Faunce T (2009). "Assessing the safety and cost-effectiveness of early nanodrugs". Journal of law and medicine. 16 (5): 822–45. PMID 19554862.

72. Davidson, Keay (2006-05-17). "FDA urged to limit nanoparticle use in cosmetics and sunscreens". San Francisco Chronicle. Retrieved 2007-04-20.

73. Benson, H., Sarveiya, V., Risk, S. and Roberts, M. S. (2005) . "Influence of anatomical site and topical formulation on skin penetration of sunscreens. Therapeutics and Clinical Risk Management, 1 3: 209–218." Retrieved 2012-04-01.

74. Howard, V. (2009). "Statement of Evidence: Particulate Emissions and Health

(An Bord Plenala, on Proposed Ringaskiddy Waste-to-Energy Facility)." Retrieved 2011-04-26.

75. "Nanomaterials EPA is Assessing". Environmental Protection Agency. Retrieved February 6, 2013.

76. Duarte, F. J. and James, R. O. (2003). "Tunable solid-state lasers incorporating dye-doped polymer-nanoparticle gain media". Opt. Lett. 28 (21): 2088–2090. Bibcode:2003OptL...28.2088D. doi:10.1364/OL.28.002088. PMID 14587824.

77. Conductive Polymer / Solvent Systems: Solutions or Dispersions?, Bernhard Wessling, 1996

78. Bigall, Nadja C.; Härtling, Thomas; Klose, Markus; Simon, Paul; Eng, Lukas M.; Eychmüller, Alexander. Monodisperse Platinum Nanospheres with Adjustable Diameters from 10 to 100 nm: Synthesis and Distinct Optical Properties. Nano Lett. [Online] 2008, 8 (12), pp 4588-4592. DOI: 10.1021/nl802901t

79. Ramirez, Esther; Eradès, Laurent; Phillippot, Karine; Lecante, Pierre; Chaudret, Bruno. Shape Control of Platinum Nanoparticles. Adv. Funct. Mater. [Online] 2007, 17, 2219-2228. DOI: 10.1002/adfm.200600633

80. PJF Harris (1986). "Sulphur-induced faceting of platinum catalyst particles.". Nature. 323 (6091): 792–94. Bibcode:1986Natur.323..792H. doi:10.1038/323792a0.

81. TS Ahmadi, ZL Wang, TC Green, A Henglein and MA El-Sayed (1996) Shape controlled synthesis of colloidal platinum nanoparticles. Science 272, 1924-1926.

82. Kim J, Takahashi M, Shimizu T, et al. (June 2008). "Effects of a potent antioxidant, platinum nanoparticle, on the lifespan of Caenorhabditis elegans". Mech. Ageing Dev. 129 (6): 322–31. doi:10.1016/j.mad.2008.02.011. PMID 18400258.

83. Meng, Hui; Zhan, Yungfeng; Zeng, Dongrong; Zhang, Xiaoxue; Zhang, Gouqing; Jaouen Frédéric. Factors Influencing the Growth of Pt Nanowires via Chemical Self-Assembly and their Fuel Cell Performance. Small [Online] 2015, Vol. 11, Issue 27, pp 3377-3386. DOI: 10.1002/smll.201402904

84. Narayanan, Radha; El-Sayed, Mostafa A.; Shape-Dependent Catalytic Activity of Platinum Nanoparticles in Colloidal Solution. Nano Lett. [Online] 2004, 4 (7), pp 1343–1348. DOI: 10.1021/nl0495256

85. Devi, G. Sarala; Rao, V. J. Room temperature synthesis of colloidal platinum nanoparticles. Bull. Mater. Sci. [Online] 2000, Vol. 23, No. 6, pp 467-470 http://www.ias.ac.in/public/Downloads/boms_023_06_0467-0470.pdf

86. Islam MA, Bhuiya MAK, Islam MS (2014). A review on chemical synthesis process of platinum nanoparticles. Asia Pacific Journal of Energy and Environment 1(2):107-20.

87. Song, Jae Yong; Kwon, Eun-Yeong; Kim, Beom Soo (2009-08-23). "Biological synthesis of platinum nanoparticles using Diopyros kaki leaf extract". Bioprocess and Biosystems Engineering. 33 (1): 159–164. doi:10.1007/s00449-009-0373-2. ISSN 1615-7591.

88. Soundarrajan, C.; Sankari, A.; Dhandapani, P.; Maruthamuthu, S.; Ravichandran, S.; Sozhan, G.; Palaniswamy, N. (2011-12-14). "Rapid biological synthesis of

platinum nanoparticles using Ocimum sanctum for water electrolysis applications". Bioprocess and Biosystems Engineering. 35 (5): 827–833. doi:10.1007/s00449-011-0666-0. ISSN 1615-7591.

89. Kharissova, Oxana V.; Dias, H. V. Rasika; Kharisov, Boris I.; Pérez, Betsabee Olvera; Pérez, Victor M. Jiménez (2013-01-04). "The greener synthesis of nanoparticles". Trends in Biotechnology. 31 (4): 240–248. doi:10.1016/j.tibtech.2013.01.003. ISSN 0167-7799.

90. Ramirez, E.; Eradès, L.; Philippot, K.; Lecante, P.; Chaudret, B. (2007-09-03). "Shape Control of Platinum Nanoparticles". Advanced Functional Materials. 17 (13): 2219–2228. doi:10.1002/adfm.200600633. ISSN 1616-3028.

91. Ahmadi, null; Wang, null; Green, null; Henglein, null; El-Sayed, null (1996-06-28). "Shape-Controlled Synthesis of Colloidal Platinum Nanoparticles". Science (New York, N.Y.). 272 (5270): 1924–1926. Bibcode:1996Sci...272.1924A. doi:10.1126/science.272.5270.1924. ISSN 1095-9203. PMID 8662492.

92. Herricks, Thurston; Chen, Jingyi; Xia, Younan (2004-12-01). "Polyol Synthesis of Platinum Nanoparticles: Control of Morphology with Sodium Nitrate". Nano Letters. 4 (12): 2367–2371. Bibcode:2004NanoL...4.2367H. doi:10.1021/nl048570a. ISSN 1530-6984.

93. Reddington E1, Sapienza A, Gurau B, Viswanathan R, Sarangapani S, Smotkin ES, Mallouk TE. Combinatorial electrochemistry: A highly parallel, optical screening method for discovery of better electrocatalysts, Science. 1998 Jun 12;280(5370):1735-7.

94. Keith R. Williams, G.Tim Burstein, Low temperature fuel cells: Interactions between catalysts and engineering design, Department of Materials Science and Metallurgy, University of Cambridge, Pembroke Street, Cambridge CB2 3QZ, UK, Catalysis Today (Impact Factor: 3.89). 11/1997; 38(4):401-410. DOI: 10.1016/S0920-5861(97)00051-5

95. Bell, A.T. (2003). The impact of nanoscience on heterogeneous catalysis. Science, 299: 1688–1691.

96. Sun S1, Murray CB, Weller D, Folks L, Moser A., Monodisperse FePt nanoparticles and ferromagnetic FePt nanocrystal superlattices, Science. 2000 Mar 17;287(5460):1989-92.

97. Chen, C., & Akashi, M. (1997). Synthesis, characterization, and catalytic properties of colloidal platinum nanoparticles protected by poly(N-isopropylacrylamide). Langmuir, 13(24), 6465-6472. doi:10.1021/la970634s

98. Katherine A. Willets and Richard P. Van Duyne, Localized Surface Plasmon Resonance Spectroscopy and Sensing. Department of Chemistry, Northwestern University, Evanston, Illinois 60208-3113

99. A.L. Stepanov, A.N. Golubev and S.I. Nikitin, Synthesis and Applications of Platinum Nanoparticles: A Review

100. Zhang, N.; Han, C.; Xu, Y.-J.; Iv, J. J. F.; Zhang, D.; Codrington, J.; Gray, S. K.; Sun, Y. Near-field dielectric scattering promotes optical absorption by platinum nanoparticles. Nature Photonics 2016, 10 (7), 473–482.

101. Shiraishi, Y.; Sakamoto, H.; Fujiwara, K.; Ichikawa, S.; Hirai, T. Pt–Cu Bimetallic Alloy Nanoparticles Supported on Anatase TiO2: Highly Active Catalysts for Aerobic Oxidation Driven by Visible Light. ACS Catalysis 2014, 4 (8), 2418–2425.

102. Mayer, K. M.; Hafner, J. H. Localized Surface Plasmon Resonance Sensors. Chemical Reviews 2011, 111 (6), 3828–3857.

103. Jain, P. K.; Huang, X.; El-Sayed, I. H.; El-Sayed, M. A. Review of Some Interesting Surface Plasmon Resonance-enhanced Properties of Noble Metal Nanoparticles and Their Applications to Biosystems. Accounts of Chemical Research Acc. Chem. Res. 2008, 41 (12), 1578–1586.

104. Chen, Aicheng; Holt-Hindle, Peter. Platinum-based nanostructured materials: synthesis, properties, and applications. Chemical reviews, 06/2010, 110 (6).

105. Repins, I., Contreras, M. A., Egaas, B., DeHart, C., Scharf, J., Perkins, C. L., ... & Noufi, R. (2008). 19· 9% efficient ZnO/CdS/CuInGaSe2 solar cell with 81· 2% fill factor. Progress in Photovoltaics: Research and applications, 16(3), 235-239.

106. Lue, J.T., Huang, W.C. and Ma, S.K. (1995). Spin–flip scattering for the electrical property of metallic–nanoparticle thin films. Phys. Rev. B, 51: 14570–14575.

107. Choi, Y.J., Park, H.H., Kim, H., Park, H.H., Chang, H.J. and Jeon, H. (2009). Fabrication and characterization of direct–patternable ZnO films containing Pt nanoparticles. Jap. J. Appl. Phys., 48: 35504–1–35504–4.

108. Gunter Oberdorster, E. O., and Jan Oberdorster, Nanotoxicology: An Emerging Discipline Evolving from Studies of Ultrafine Particles. Environmental Health Perspectives 2005, 113 (7), 823-839

109. Pelka, J.; Gehrke, H.; Esselen, M.; Turk, M.; Crone, M.; Brase, S.; Muller, T.; Blank, H.; Send, W.; Zibat, V.; Brenner, P.; Schneider, R.; Gerthsen, D.; Marko, D., Cellular uptake of platinum nanoparticles in human colon carcinoma cells and their impact on cellular redox systems and DNA integrity. Chem Res Toxicol 2009, 22.

110. Dhar, S.; Gu, F. X.; Langer, R.; Farokhzad, O. C.; Lippard, S. J., Targeted delivery of cisplatin to prostate cancer cells by aptamer functionalized Pt(IV) prodrug-PLGA–PEG nanoparticles. Proceedings of the National Academy of Sciences 2008, 105 (45), 17356-17361

111. Konieczny, P; Goralczyk, A; Szmyd, R; Skalniak, L; Koziel, J; Filon, F; Crosera, M, Cierniak, A; Zuba-Surma, E; Borowczyk, J.; Effects triggered by platinum nanoparticles on primary keratinocytes. International Journal of Nanomedicine. 2013. 8(1), 3963-3975 DOI: 10.2147/IJN.S49612.

CHAPTER 3

Nanoelectronics

Chapter Contents

- **Introduction**
- **Fundamental Concepts**
- **Approaches To Nanoelectronics**
- **Nanoelectronic Devices**
- **Molecular Logic Gate**
- **Nanocircuitry**
- **Molecular Wires**
- **Nanosensors**
- **Nanoionics**

Nanoelectronics is defined as nanotechnology which allows the integration of purely electronic devices, electronic chips and circuits. The digital systems are combined with analog/RF circuits. This type of technology fusion can be described as the 'More than Moore' domain of development. The nanoscale dimensions of nanoelectronic components for systems of giga-scale complexity measured on a chip or in a package. This scaling feature and the road to giga-scale systems can be described as the 'More Moore' domain of development.

INTRODUCTION

Nanoelectronics refer to the use of nanotechnology in electronic components. The term covers a diverse set of devices and materials, with the common characteristic that they are so small that inter-atomic interactions and quantum mechanical properties need to be studied extensively. Some of these candidates include: hybrid molecular/semiconductor electronics, one-dimensional nanotubes/nanowires, or advanced molecular electronics. Recent silicon CMOS technology generations, such as the 22 nanometer node, are already within this regime. Nanoelectronics are sometimes considered as disruptive technology because present candidates are significantly different from traditional transistors.

FUNDAMENTAL CONCEPTS

In 1965 Gordon Moore observed that silicon transistors were undergoing a continual process of scaling downward, an observation which was later codified as Moore's law. Since his observation transistor minimum feature sizes have decreased from 10 micrometers to the 28-22 nm range in 2011. The field of nanoelectronics aims to enable the continued realization of this law by using new methods and materials to build electronic devices with feature sizes on the nanoscale. The volume of an object decreases as the third power of its linear dimensions, but the surface area only decreases as its second power. This somewhat subtle and unavoidable principle has huge ramifications. For example, the power of a drill (or any other machine) is proportional to the volume, while the friction of the drill's bearings and gears is proportional to their surface area. For a normal-sized drill, the power of the device is enough to handily overcome any friction. However, scaling its length down by a factor of 1000, for example, decreases its power by 1000^3 (a factor of a billion) while reducing the friction by only 1000^2 (a factor of only a million). Proportionally it has 1000 times less power per unit friction than the original drill. If the original friction-to-power ratio was, say, 1%, that implies the smaller drill will have 10 times as much friction as power; the drill is useless. For this reason, while super-miniature electronic integrated circuits are fully functional, the same technology cannot be used to make working mechanical devices beyond the scales where frictional forces start to exceed the available power. So even though you may see microphotographs of delicately etched silicon gears, such devices are currently little more than curiosities with limited real world applications, for example, in moving mirrors and shutters. Surface tension increases in much the same way, thus magnifying the tendency for very small objects to stick together. This could possibly make any kind of "micro factory" impractical: even if robotic arms and hands could be scaled down, anything they pick up will tend to be impossible to put down. The above being said, molecular evolution has resulted in working cilia, flagella, muscle fibers and rotary motors in aqueous environments, all on the nanoscale. These machines exploit the increased frictional forces found at the micro or nanoscale. Unlike a paddle or a propeller which depends on normal frictional forces (the frictional forces perpendicular to the surface) to achieve propulsion, cilia develop motion from the exaggerated drag or laminar forces (frictional forces parallel to the surface) present at micro and nano dimensions. To build meaningful "machines" at the nanoscale, the relevant forces need to be considered. We are faced with the development and design of intrinsically pertinent machines rather than the simple reproductions of macroscopic ones. All scaling issues therefore need to be assessed thoroughly when evaluating nanotechnology for practical applications.

APPROACHES TO NANOELECTRONICS

Nanofabrication

For example, single electron transistors, which involve transistor operation based on a single electron. Nanoelectromechanical systems also fall under this category. Nanofabrication can be used to construct ultradense parallel arrays of nanowires, as an alternative to synthesizing nanowires individually.

Nanomaterials Electronics

Besides being small and allowing more transistors to be packed into a single chip, the uniform and symmetrical structure of nanotubes allows a higher electron mobility (faster electron movement in the material), a higher dielectric constant (faster frequency), and a symmetrical electron/hole characteristic.

Also, nanoparticles can be used as quantum dots.

Molecular Electronics

Molecular scale electronics, also called single-molecule electronics, is a branch of nanotechnology that uses single molecules, or nanoscale collections of single molecules, as electronic components. Because single molecules constitute the smallest stable structures imaginable, this miniaturization is the ultimate goal for shrinking electrical circuits.

The field is often termed simply as "molecular electronics", but this term is also used to refer to the distantly related field of conductive polymers and organic electronics, which uses the properties of molecules to affect the bulk properties of a material. A nomenclature distinction has been suggested so that molecular materials for electronics refers to this latter field of bulk applications, while molecular scale electronics refers to the nanoscale single-molecule applications treated here.

Fundamental Concepts

Conventional electronics have traditionally been made from bulk materials. Ever since their invention in 1958, the performance and complexity of integrated circuits has undergone exponential growth, a trend named Moore's law, as feature sizes of the embedded components have shrunk accordingly. As the structures shrink, the sensitivity to deviations increases. In a few technology generations, when the minimum feature sizes reaches 13 nm, the composition of the devices must be controlled to a precision of a few atoms for the devices to work. With bulk methods growing increasingly demanding and costly as they near inherent limits, the idea was born that the components could instead be built up atom by atom in a chemistry lab (bottom up) versus carving them out of bulk material (top down). This is the idea behind molecular electronics, with the ultimate miniaturization being components contained in single molecules.

In single-molecule electronics, the bulk material is replaced by single molecules. Instead of forming structures by removing or applying material after a pattern scaffold, the atoms are put together in a chemistry lab. In this way, billions of billions of copies are made simultaneously (typically more than 1020 molecules are made at once) while the composition of molecules are controlled down to the last atom. The molecules used have properties that resemble traditional electronic components such as a wire, transistor or rectifier. Single-molecule electronics is an emerging field, and entire electronic circuits consisting exclusively of molecular sized compounds are still very far from being realized. However, the unceasing demand for more computing power, along with the inherent limits of lithographic methods as of 2016, make the transition seem unavoidable. Currently, the focus is on discovering molecules with interesting properties and on finding ways to obtain reliable and reproducible contacts between the molecular components and the bulk material of the electrodes.

Theoretical Basis

Molecular electronics operates in the quantum realm of distances less than 100 nanometers. The miniaturization down to single molecules brings the scale down to a regime where quantum mechanics effects are important. In conventional electronic components, electrons can be filled in or drawn out more or less like a continuous flow of electric charge. In contrast, in molecular electronics the transfer of one electron alters the system significantly. For example, when an electron has been transferred from a source electrode to a molecule, the molecule gets charged up, which makes it far harder for the next electron to transfer (see also Coulomb blockade). The significant amount of energy due to charging must be accounted for when making calculations about the electronic properties of the setup, and is highly sensitive to distances to conducting surfaces nearby.

The theory of single-molecule devices is especially interesting since the system under consideration is an open quantum system in nonequilibrium (driven by voltage). In the low bias voltage regime, the nonequilibrium nature of the molecular junction can be ignored, and the current-voltage traits of the device can be calculated using the equilibrium electronic structure of the system. However, in stronger bias regimes a more sophisticated treatment is required, as there is no longer a variational principle. In the elastic tunneling case (where the passing electron does not exchange energy with the system), the formalism of Rolf Landauer can be used to calculate the transmission through the system as a function of bias voltage, and hence the current. In inelastic tunneling, an elegant formalism based on the non-equilibrium Green's functions of Leo Kadanoff and Gordon Baym, and independently by Leonid Keldysh was advanced by Ned Wingreen and Yigal Meir. This Meir-Wingreen formulation has been used to great success in the molecular electronics community to examine the more difficult and interesting cases where the transient electron exchanges energy with the molecular system (for example through electron-phonon coupling or electronic excitations).

Further, connecting single molecules reliably to a larger scale circuit has proven a great challenge, and constitutes a significant hindrance to commercialization.

Examples

Common for molecules used in molecular electronics is that the structures contain many alternating double and single bonds (see also Conjugated system). This is done because such patterns delocalize the molecular orbitals, making it possible for electrons to move freely over the conjugated area.

Wires

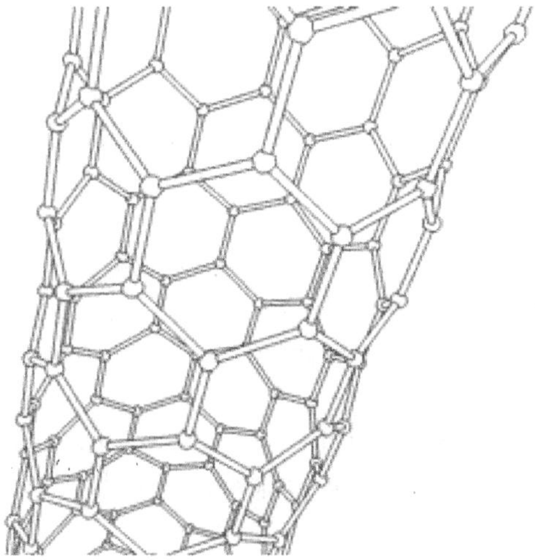

This animation of a rotating carbon nanotube shows its 3D structure.

The sole purpose of molecular wires is to electrically connect different parts of a molecular electrical circuit. As the assembly of these and their connection to a macroscopic circuit is still not mastered, the focus of research in single-molecule electronics is primarily on the functionalized molecules: molecular wires are characterized by containing no functional groups and are hence composed of plain repetitions of a conjugated building block. Among these are the carbon nanotubes that are quite large compared to the other suggestions but have shown very promising electrical properties.

The main problem with the molecular wires is to obtain good electrical contact with the electrodes so that electrons can move freely in and out of the wire.

Transistors

Single-molecule transistors are fundamentally different from the ones known from bulk electronics. The gate in a conventional (field-effect) transistor determines the conductance between the source and drain electrode by controlling the density of charge carriers between them, whereas the gate in a single-molecule transistor controls the possibility of a single electron to jump on and off the molecule by modifying the energy of the molecular orbitals. One of the effects of this difference is that the single-molecule transistor is almost binary: it is either on or off. This opposes its bulk counterparts, which have quadratic responses to gate voltage. It is the quantization of charge into electrons that is responsible for the markedly different behavior compared to bulk electronics. Because of the size of a single molecule, the charging due to a single electron is significant and provides means to turn a transistor on or off (see Coulomb blockade). For this to work, the electronic orbitals on the transistor molecule cannot be too well integrated with the orbitals on the electrodes. If they are, an electron cannot be said to be located on the molecule or the electrodes and the molecule will function as a wire.

A popular group of molecules, that can work as the semiconducting channel material in a molecular transistor, is the oligopolyphenylenevinylenes (OPVs) that works by the Coulomb blockade mechanism when placed between the source and drain electrode in an appropriate way. Fullerenes work by the same mechanism and have also been commonly used.

Semiconducting carbon nanotubes have also been demonstrated to work as channel material but although molecular, these molecules are sufficiently large to behave almost as bulk semiconductors.

The size of the molecules, and the low temperature of the measurements being conducted, makes the quantum mechanical states well defined. Thus, it is being researched if the quantum mechanical properties can be used for more advanced purposes than simple transistors (e.g. spintronics).

Physicists at the University of Arizona, in collaboration with chemists from the University of Madrid, have designed a single-molecule transistor using a ring-shaped molecule similar to benzene. Physicists at Canada's National Institute for Nanotechnology have designed a single-molecule transistor using styrene. Both groups expect (the designs were experimentally unverified as of June 2005) their respective devices to function at room temperature, and to be controlled by a single electron.

Rectifiers (diodes)

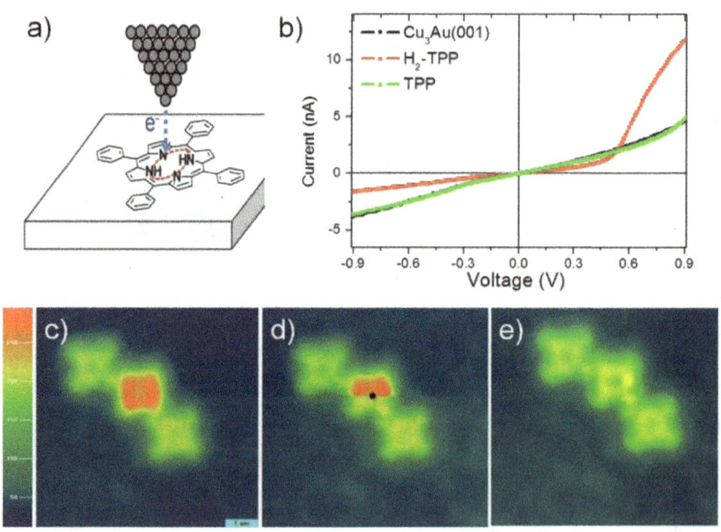

Hydrogen can be removed from individual tetraphenylporphyrin (H_2TPP) molecules by applying excess voltage to the tip of a scanning tunneling microscope (STAM, a); this removal alters the current-voltage (I-V) curves of TPP molecules, measured using the same STM tip, from diode-like (red curve in b) to resistor-like (green curve). Image (c) shows a row of TPP, H_2TPP and TPP molecules. While scanning image (d), excess voltage was applied to H_2TPP at the black dot, which instantly removed hydrogen, as shown in the bottom part of (d) and in the re-scan image (e). Such manipulations can be used in single-molecule electronics.

Molecular rectifiers are mimics of their bulk counterparts and have an asymmetric construction so that the molecule can accept electrons in one end but not the other. The molecules have an electron donor (D) in one end and an electron acceptor (A) in the other. This way, the unstable state D+ – A− will be more readily made than D− – A+. The result is that an electric current can be drawn through the molecule if the electrons are added through the acceptor end, but less easily if the reverse is attempted.

Methods

One of the biggest problems with measuring on single molecules is to establish reproducible electrical contact with only one molecule and doing so without shortcutting the electrodes. Because the current photolithographic technology is unable to produce electrode gaps small enough to contact both ends of the molecules tested (on the order of nanometers), alternative strategies are applied.

Molecular Gaps

One way to produce electrodes with a molecular sized gap between them is break junctions, in which a thin electrode is stretched until it breaks. Another is electromigration. Here a current is led through a thin wire until it melts and the atoms migrate to produce the gap. Further, the reach of conventional photolithography can be enhanced by chemically etching or depositing metal on the electrodes.

Probably the easiest way to conduct measurements on several molecules is to use the tip of a scanning tunneling microscope (STM) to contact molecules adhered at the other end to a metal substrate.

Anchoring

A popular way to anchor molecules to the electrodes is to make use of sulfur's high chemical affinity to gold. In these setups, the molecules are synthesized so that sulfur atoms are placed strategically to function as crocodile clips connecting the molecules to the gold electrodes. Though useful, the anchoring is non-specific and thus anchors the molecules randomly to all gold surfaces. Further, the contact resistance is highly dependent on the precise atomic geometry around the site of anchoring and thereby inherently compromises the reproducibility of the connection.

To circumvent the latter issue, experiments has shown that fullerenes could be a good candidate for use instead of sulfur because of the large conjugated π-system that can electrically contact many more atoms at once than one atom of sulfur.

Fullerene Nanoelectronics

In polymers, classical organic molecules are composed of both carbon and hydrogen (and sometimes additional compounds such as nitrogen, chlorine or sulphur). They are obtained from petrol and can often be synthesized in large amounts. Most of these molecules are insulating when their length exceeds a few nanometers. However, naturally occurring carbon is conducting, especially graphite recovered from coal or encountered otherwise.

From a theoretical viewpoint, graphite is a semi-metal, a category in between metals and semi-conductors. It has a layered structure, each sheet being one atom thick. Between each sheet, the interactions are weak enough to allow an easy manual cleavage.

Tailoring the graphite sheet to obtain well defined nanometer-sized objects remains a challenge. However, by the close of the twentieth century, chemists were exploring methods to fabricate extremely small graphitic objects that could be considered single molecules. After studying the interstellar conditions under which carbon is known to form clusters, Richard Smalley's group (Rice University, Texas) set up an experiment in which graphite was vaporized via laser irradiation. Mass spectrometry revealed that clusters containing specific magic numbers of atoms were stable, especially those clusters of 60 atoms. Harry Kroto, an English chemist who assisted in the experiment, suggested a possible geometry for these clusters – atoms covalently bound with the exact symmetry of a soccer ball. Coined buckminsterfullerenes, buckyballs, or C60, the clusters retained some properties of graphite, such as conductivity. These objects were rapidly envisioned as possible building blocks for molecular electronics.

Problems

Artifacts

When trying to measure electronic traits of molecules, artificial phenomena can occur that can be hard to distinguish from truly molecular behavior. Before they were discovered, these artifacts have mistakenly been published as being features pertaining to the molecules in question. Applying a voltage drop on the order of volts across a nanometer sized junction results in a very strong electrical field. The field can cause metal atoms to migrate and eventually close the gap by a thin filament, which can be broken again when carrying a current. The two levels of conductance imitate molecular switching between a conductive and an isolating state of a molecule.

Another encountered artifact is when the electrodes undergo chemical reactions due to the high field strength in the gap. When the voltage bias is reversed, the reaction will cause hysteresis in the measurements that can be interpreted as being of molecular origin.

A metallic grain between the electrodes can act as a single electron transistor by the mechanism described above, thus resembling the traits of a molecular transistor. This artifact is especially common with nanogaps produced by the electromigration method.

Commercialization

One of the biggest hindrances for single-molecule electronics to be commercially exploited is the lack of methods to connect a molecular sized circuit to bulk electrodes in a way that gives reproducible results. At the current state, the difficulty of connecting single molecules vastly outweighs any possible performance increase that could be gained from such shrinkage. The difficulties grow worse if the molecules are to have a certain spatial orientation and/or have multiple poles to connect.

Also problematic is that some measurements on single molecules are carried out in cryogenic temperatures (near absolute zero), which is very energy consuming. This is done to reduce signal noise enough to measure the faint currents of single molecules.

History and Recent Progress

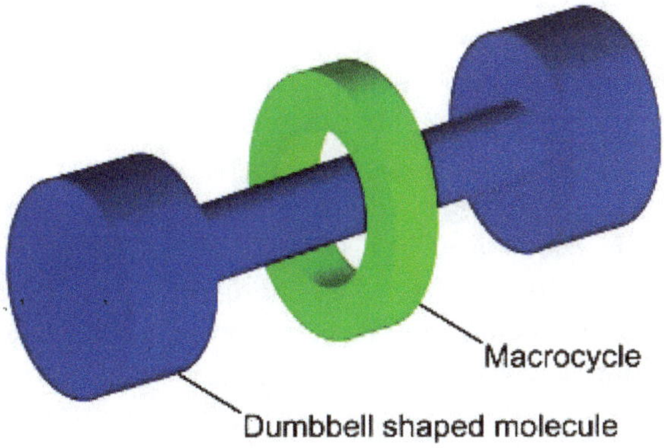

Graphical representation of a rotaxane, useful as a molecular switch.

In their treatment of so-called donor-acceptor complexes in the 1940s, Robert Mulliken and Albert Szent-Györgyi advanced the concept of charge transfer in molecules. They subsequently further refined the study of both charge transfer and energy transfer in molecules. Likewise, a 1974 paper from Mark Ratner and Ari Aviram illustrated a theoretical molecular rectifier. In 1988, Aviram described in detail a theoretical single-molecule field-effect transistor. Further concepts were proposed by Forrest Carter of the Naval Research Laboratory, including single-molecule logic gates. A wide range of ideas were presented, under his aegis, at a conference entitled Molecular Electronic Devices in 1988. These were all theoretical constructs and not concrete devices. The direct measurement of the electronic traits of individual molecules awaited the development of methods for making molecular-scale electrical contacts. This was no easy task. Thus, the first experiment directly-measuring the conductance of a single molecule was only reported in 1995 on a single C60molecule by C. Joachim and J. K. Gimzewsky in their seminal Physical Revie Letter paper and later in 1997 by Mark Reed and co-workers on a few hundred molecules. Since then, this branch of the field has advanced rapidly. Likewise, as it has grown possible to measure such properties directly, the theoretical predictions of the early workers have been confirmed substantially.

Recent progress in nanotechnology and nanoscience has facilitated both experimental and theoretical study of molecular electronics. Development of the scanning tunneling microscope (STM) and later the atomic force microscope (AFM) have greatly facilitated manipulating single-molecule electronics. Also, theoretical advances in molecular electronics have facilitated further understanding of non-adiabatic charge transfer events at electrode-electrolyte interfaces.

The concept of molecular electronics was first published in 1974 when Aviram and Ratner suggested an organic molecule that could work as a rectifier. Having both huge commercial and fundamental interest, much effort was put into proving its feasibility, and 16 years later in 1990, the first demonstration of an intrinsic molecular rectifier was realized by Ashwell and coworkers for a thin film of molecules.

The first measurement of the conductance of a single molecule was realised in 1994 by C. Joachim and J. K. Gimzewski and published in 1995 (see the corresponding Phys. Rev. Lett. paper). This was the conclusion of 10 years of research started at IBM TJ Watson, using the scanning tunnelling microscope tip apex to switch a single molecule as already explored by A. Aviram, C. Joachim and M. Pomerantz at the end of the 80's (see their seminal Chem. Phys. Lett. paper during this period). The trick was to use an UHV Scanning Tunneling microscope to allow the tip apex to gently touch the top of a single C60 molecule adsorbed on an Au(110) surface. A resistance of 55 MOhms was recorded along with a low voltage linear I-V. The contact was certified by recording the I-z current distance property, which allows measurement of the deformation of the C60 cage under contact. This first experiment was followed by the reported result using a mechanical break junction method to connect two gold electrodes to a sulfur-terminated molecular wire by Mark Reed and James Tour in 1997. A single-molecule amplifier was implemented by C. Joachim and J.K. Gimzewski in IBM Zurich. This experiment, involving one C60 molecule, demonstrated that one such molecule can provide gain in a circuit via intramolecular quantum interference effects alone. A collaboration of researchers at Hewlett-Packard (HP) and University of California, Los Angeles (UCLA), led by James Heath, Fraser Stoddart, R. Stanley Williams, and Philip Kuekes, has developed molecular electronics based on rotaxanes and catenanes. Work is also occurring on the use of single-wall carbon nanotubes as field-effect transistors. Most of this work is being done by International Business Machines (IBM). Some specific reports of a field-effect transistor based on molecular self-assembled monolayers were shown to be fraudulent in 2002 as part of the Schön scandal. Until recently entirely theoretical, the Aviram-Ratner model for a unimolecular rectifier has been confirmed unambiguously in experiments by a group led by Geoffrey J. Ashwell at Bangor University, UK. Many rectifying molecules have so far been identified, and the number and efficiency of these systems is growing rapidly.

Supramolecular electronics is a new field involving electronics at a supramolecular level. An important issue in molecular electronics is the determination of the resistance of a single molecule (both theoretical and experimental). For example, Bumm, et al. used STM to analyze a single molecular switch in a self-assembled monolayer to determine how conductive such a molecule can be. Another problem faced by this field is the difficulty of performing direct characterization since imaging at the molecular scale is often difficult in many experimental devices.

Other Approaches

Nanoionics studies the transport of ions rather than electrons in nanoscale systems.

Nanophotonics studies the behavior of light on the nanoscale, and has the goal of developing devices that take advantage of this behavior.

NANOELECTRONIC DEVICES

Current high-technology production processes are based on traditional top down strategies, where nanotechnology has already been introduced silently. The critical length scale of integrated circuits is already at the nanoscale (50 nm and below) regarding the gate length of transistors in CPUs or DRAM devices.

Computers

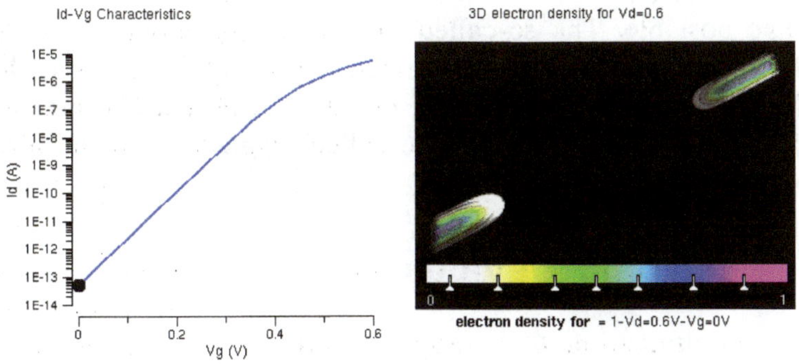

Simulation result for formation of inversion channel (electron density) and attainment of threshold voltage (IV) in a nanowire MOSFET. Note that the threshold voltage for this device lies around 0.45V.

Nanoelectronics holds the promise of making computer processors more powerful than are possible with conventional semiconductor fabrication techniques. A number of approaches are currently being researched, including new forms of nanolithography, as well as the use of nanomaterials such as nanowires or small molecules in place of traditional CMOS components. Field effect transistors have been made using both semiconducting carbon nanotubes and with heterostructured semiconductor nanowires.

In 1999, the CMOS transistor developed at the Laboratory for Electronics and Information Technology in Grenoble, France, tested the limits of the principles of the MOSFET transistor with a diameter of 18 nm (approximately 70 atoms placed side by side). This was almost one tenth the size of the smallest industrial transistor in 2003 (130 nm in 2003, 90 nm in 2004, 65 nm in 2005 and 45 nm in 2007). It enabled the theoretical integration of seven billion junctions on a €1 coin. However, the CMOS transistor, which was created in 1999, was not a simple research experiment to study how CMOS technology functions, but rather a demonstration of how this technology functions now that we ourselves are getting ever closer to working on a molecular scale. Today it would be impossible to master the coordinated assembly of a large number of these transistors on a circuit and it would also be impossible to create this on an industrial level.

Memory Storage

Electronic memory designs in the past have largely relied on the formation of transistors. However, research into crossbar switch based electronics have offered an alternative using reconfigurable interconnections between vertical and horizontal wiring arrays to create ultra high density memories. Two leaders in this area are Nantero which has developed a carbon nanotube based crossbar memory called Nano-RAM and Hewlett-Packard which has proposed the use of memristor material as a future replacement of Flash memory.

An example of such novel devices is based on spintronics.The dependence of the resistance of a material (due to the spin of the electrons) on an external field is called magnetoresistance. This effect can be significantly amplified (GMR - Giant Magneto-Resistance) for nanosized objects, for example when two ferromagnetic layers are separated by a nonmagnetic layer, which is several nanometers thick (e.g. Co-Cu-Co). The GMR effect has led to a strong increase in the data storage density of hard disks and made the gigabyte range possible. The so-called tunneling magnetoresistance (TMR) is very similar to GMR and based on the spin dependent tunneling of electrons through adjacent ferromagnetic layers. Both GMR and TMR effects can be used to create a non-volatile main memory for computers, such as the so-called magnetic random access memory or MRAM.

Novel Optoelectronic Devices

In the modern communication technology traditional analog electrical devices are increasingly replaced by optical or optoelectronic devices due to their enormous bandwidth and capacity, respectively. Two promising examples are photonic crystals and quantum dots. Photonic crystals are materials with a periodic variation in the refractive index with a lattice constant that is half the wavelength of the light used. They offer a selectable band gap for the propagation of a certain wavelength, thus they resemble a semiconductor, but for light or photons instead of electrons. Quantum dots are nanoscaled objects, which can be used, among many other things, for the construction of lasers. The advantage of a quantum dot laser over the traditional semiconductor laser is that their emitted wavelength depends on the diameter of the dot. Quantum dot lasers are cheaper and offer a higher beam quality than conventional laser diodes.

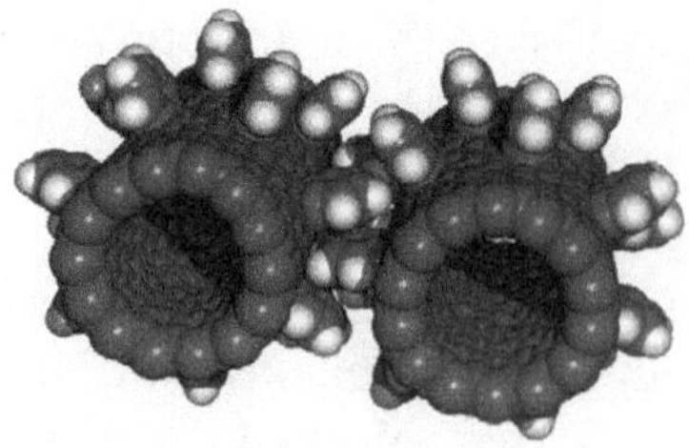

Displays

The production of displays with low energy consumption might be accomplished using carbon nanotubes (CNT). Carbon nanotubes are electrically conductive and due to their small diameter of several nanometers, they can be used as field emitters with extremely high efficiency for field emission displays (FED). The principle of operation resembles that of the cathode ray tube, but on a much smaller length scale.

Quantum Computers

Entirely new approaches for computing exploit the laws of quantum mechanics for novel quantum computers, which enable the use of fast quantum algorithms. The Quantum computer has quantum bit memory space termed "Qubit" for several computations at the same time. This facility may improve the performance of the older systems.

Radios

Nanoradios have been developed structured around carbon nanotubes.

Energy Production

Research is ongoing to use nanowires and other nanostructured materials with the hope to create cheaper and more efficient solar cells than are possible with conventional planar silicon solar cells. It is believed that the invention of more efficient solar energy would have a great effect on satisfying global energy needs. There is also research into energy production for devices that would operate *in vivo*, called bio-nano generators. A bio-nano generator is a nanoscale electrochemical device, like a fuel cell or galvanic cell, but drawing power from blood glucose in a living body, much the same as how the body generates energy from food. To achieve the effect, an enzyme is used that is capable of stripping glucose of its electrons, freeing them for use in electrical devices. The average person's body could, theoretically, generate 100 watts of electricity (about 2000 food calories per day) using a bio-nano generator. However, this estimate is only true if all food was converted to electricity, and the human body needs some energy consistently, so possible power generated is likely much lower. The electricity generated by such a device could power devices embedded in the body (such as pacemakers), or sugar-fed nanorobots. Much of the research done on bio-nano generators is still experimental, with Panasonic's Nanotechnology Research Laboratory among those at the forefront.

Medical Diagnostics

There is great interest in constructing nanoelectronic devices that could detect the concentrations of biomolecules in real time for use as medical diagnostics, thus falling into the category of nanomedicine. A parallel line of research seeks to create nanoelectronic devices which could interact with single cells for use in basic biological research. These devices are called nanosensors. Such miniaturization on nanoelectronics towards in vivo proteomic sensing should enable new approaches for health monitoring, surveillance, and defense technology.

MOLECULAR LOGIC GATE

A **molecular logic gate** is a molecule that performs a logical operation based on one or more physical or chemical inputs and a single output. The field has advanced from simple logic systems based on a single chemical or physical input to molecules capable of combinatorial and sequential operations such as arithmetic operations i.e. **moleculators** and memory storage algorithms.

For logic gates with a single input, there are four possible output patterns. When the input is 0, the output can be either a 0 or 1. When the input is 1, the output can again be 0 or 1. The four output bit patterns that can arise corresponds to a specific logic type: PASS 0, YES, NOT and PASS 1. PASS 0 always outputs 0, whatever the input. PASS 1 always outputs 1, whatever the input. YES outputs a 1 when the input is 1 and NOT is the inverse YES - it outputs a 0 when the input is 1. An example of a YES logic gate is the molecular structure shown below. A '1' output is given only when sodium ions are present in solution ('1' input).

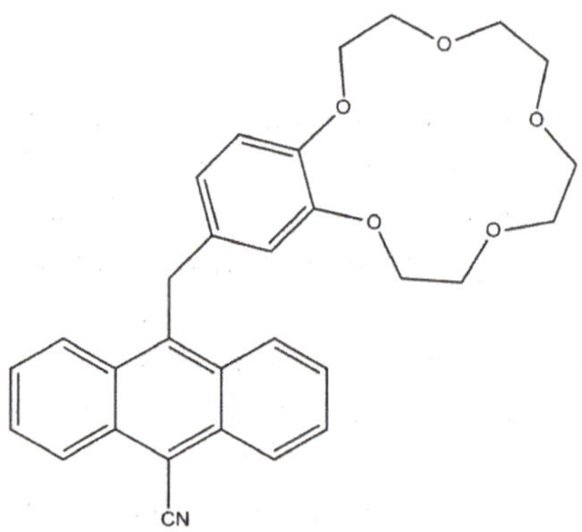

Molecular logic gates work with input signals based on chemical processes and with output signals based on spectroscopy. One of the earlier water solution-based systems exploits the chemical behavior of compounds **A** and **B** in *scheme 1* .

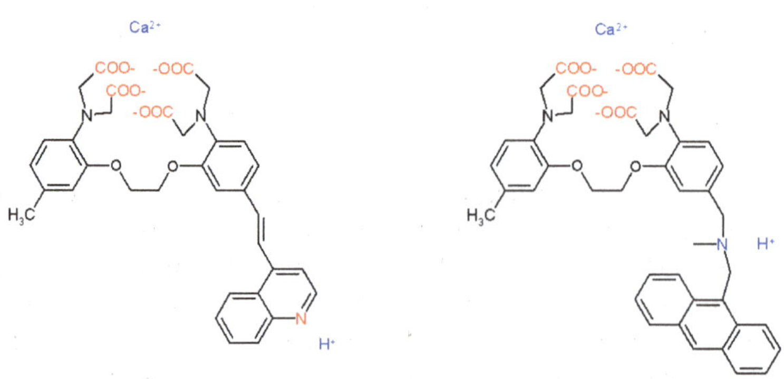

A B

Compound A is a push-pull olefin with the top receptor containing four carboxylic acid anion groups (and non-disclosed counter cations) capable of binding to calcium. The bottom part is a quinoline molecule which is a receptor for hydrogen ions. The logic gate operates as follows. Without any chemical input of Ca^{2+} or H^+, the chromophore shows a maximum absorbance in UV/VIS spectroscopy at 390 nm. When calcium is introduced a blue shift takes place and the absorbance at 390 nm decreases. Likewise addition of protons causes a red shift and when both cations are in the water the net result is absorption at the original 390 nm. This system represents a XNOR logic gate in absorption and a XOR logic gate in transmittance.

In compound **B** the bottom section now contains a tertiary amino group also capable of binding to protons. In this system fluorescence only takes place when both cations available. The presence of both cations hinders photoinduced electron transfer (PET) allowing compound B to fluoresce. In the absence of both or either ion, fluorescence is quenched by PET, which involves an electron transfer from either the nitrogen atom or the oxygen atoms, or both to the anthracenyl group. When both receptors are bound to calcium ions and protons respectively, both PET channels are shut off. The overall result of Compound B is AND logic, since an output of "1" (fluorescence) occurs only when both Ca^{2+} and H^+ are present in solution, that is, have values as "1". With both systems run in parallel and with monitoring of transmittance for system A and fluorescence for system B the result is a half-adder capable of reproducing the equation 1+1=2.

In a modification of system B not two but three chemical inputs are simultaneously processed in an AND logic gate . An enhanced fluorescence signal is observed only in the presence of excess protons, zinc and sodium ions through interactions with their respective amine, phenyldiaminocarboxylate and crown ether receptors. The processing mode operates similarly as discussed above - fluorescence is observed due to the prevention of competing photoinduced electron transfer reactions from the receptors to the excited anthracene fluorophore. The absence of one, two or all three ion inputs results in a low fluorescence output. Each receptor is selective for its specific ion as an increase in the concentration of the other ions does not yield a high fluorescence. The specific concentration threshold of each input must be reached to achieve a fluorescent output in accordance with combinatorial AND logic. This prototype could potentially be extended to point-of-care medical diagnostics application for disease screening in the future.

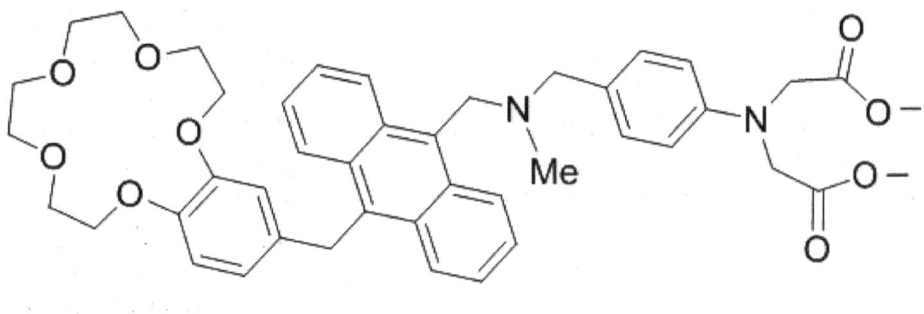

In a similar set-up, the molecular logic gate illustrated below demonstrates the advancement from redox-fluorescent switches to multi-input logic gates with an electrochemical switch. This two-input AND logic gate incorporates a tertiary amine proton receptor and a tetrathiafulvelene redox donor. These groups, when attached to anthracene can simultaneously process information concerning acid concentration and oxidizing ability of the solution.

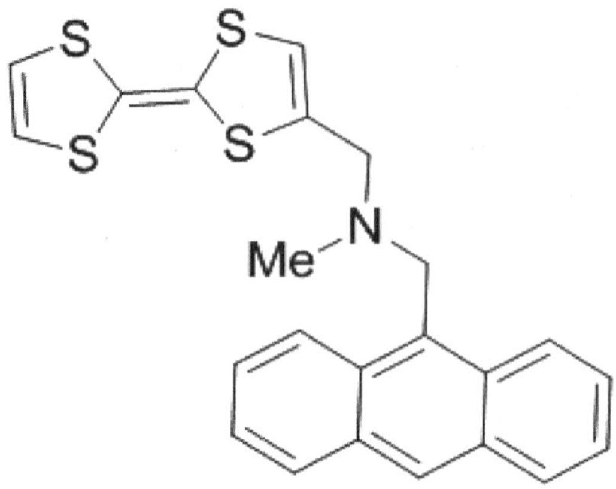

The INHIBIT logic gate illustrated below as provided by Gunnlaugsson et al. incorporates a Tb^{3+} ion in a chelate complex.This two-input logic gate is the first of its kind and displays non-commutative behaviour with chemical inputs and a phosphorescence output. Whenever dioxygen (input 1) is present, the system is quenched and no phosphorescence is observed (output 0). The second input, H^+, must also be present for an output "1" to be observed. This is understood from a two-input INHIBIT truth table.

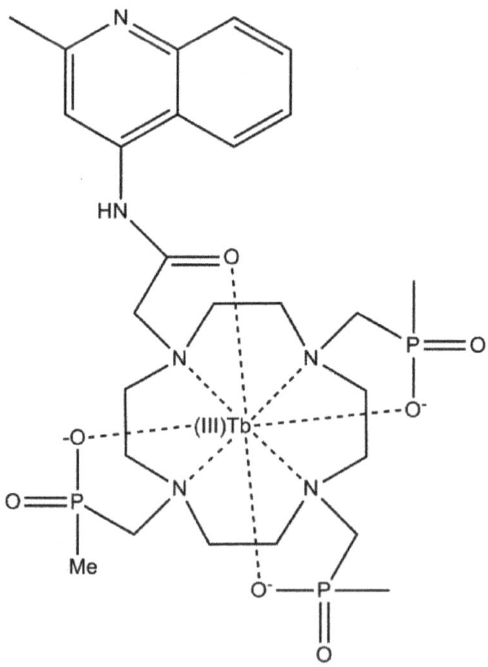

In another XOR logic gate system the chemistry is based on the pseudorotaxane depicted in *scheme 3*. In organic solution the electron deficient diazapyrenium salt (rod) and the electron rich 2,3-dioxynaphthalene units of the crown ether (ring) self-assemble by formation of a charge transfer complex.

An added tertiary amine like tributylamine forms a 1:2 adduct with the diazapyrene and the complex gets dethreaded. This process is accompanied by an increase in emission intensity at 343 nm resulting from freed crown ether. Added trifluoromethanesulfonic acid reacts with the amine and the process is reverted. Excess acid locks the crown ether by protonation and again the complex is dethreaded.

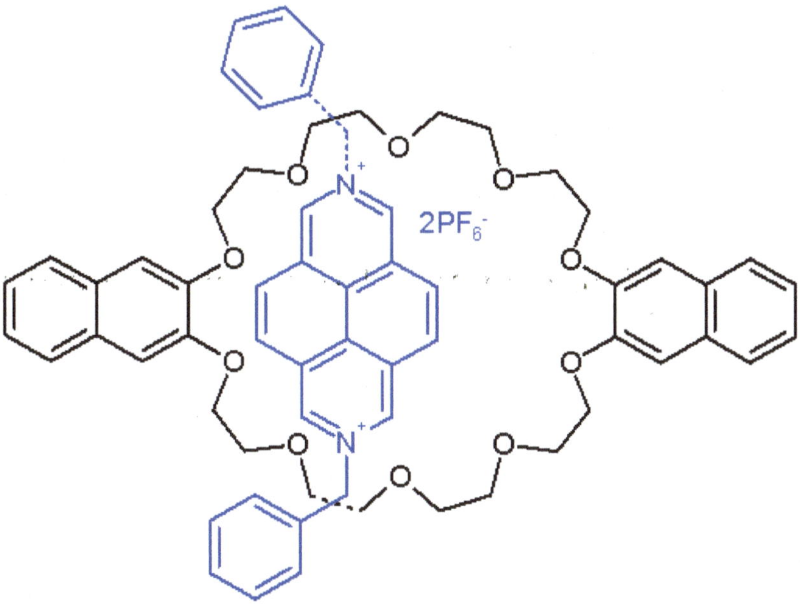

A full adder system based on fluorescein is able to compute 1+1+1=3.

Molecular sequential logic is exemplified by D. Margulies et al., where they demonstrate a molecular keypad lock resembling the processing capabilities of an electronic security device which is equivalent to incorporates several interconnected AND logic gates in parallel. The molecule mimics an electronic keypad of an automated teller machine (ATM). The output signals are dependent not only on the combination of inputs but also on the correct order of inputs: in other words the correct password must be entered. The molecule was designed using pyrene and fluorescein fluorophores connected by a siderophore, which binds to Fe(III), and the acidic of the solution changes the fluorescence properties of the fluorescein fluorophore. Further development in this field might also see molecular logic gates replace semiconductors in the IT industry. Such molecular systems can theoretically overcome the problems arising when semiconductors approach nano-dimensions. Molecular logic gates are more versatile than their silicon counterparts, with phenomena such as superposed logic unavailable to semiconductor electronics. Dry molecular gates such as the one demonstrated by Avouris and colleagues prove to be possible substitutes for semiconductor devices due to their small size, similar infrastructure and data processing abilities. Avouris revealed a NOT logic gate composed of a bundle of carbon nanotubes. The nanotubes are doped differently in adjoining regions creating two complementary field effect transistors. The bundle operates as a NOT logic gate only when satisfactory conditions are met.

New potential applications of chemical logic gates continue to be explored. A recent study illustrates the application of a logic gate for photodynamic therapy. A bodipy dye attached to a crown-ether and two pyridyl groups separated by spacers (as shown below) works according to an AND logic gate. The molecule works as a photodynamic agent upon irradiation at 660 nm under conditions of relatively high sodium and proton ion concentrations by converting triplet oxygen to cytotoxic singlet oxygen. This prototypical example would take advantage of the higher sodium levels and lower pH in tumor tissue compared to the levels in normal cells. When these two cancer-related cellular parameters are satisfied, a change is observed in the absorbance spectrum. This technique could be useful for the treatment of malignant tumors as it is non-invasive and specific.

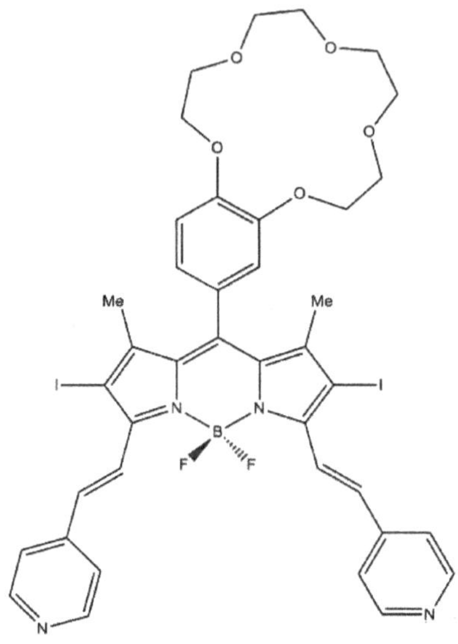

A molecular logic gate can processes modulators much like the set up seen in de Silva's 'Proof-of-principle' but incorporating different logic gates on the same molecule. Such a function is called integrated logic and is exemplified by the BODIPY-based, half-subtractor logic gate illustrated by A. Coskun, E. U. Akkaya and their colleagues (as shown below). When monitored at two different wavelengths, 565 and 660 nm, XOR and INHIBIT logic gates are obtained at the respective wavelengths. Optical studies of this compound in THF reveal an absorbance peak at 565 nm and an emission peak at 660 nm. Addition of an acid results in a hypsochromic shift of both peaks as protonation of the tertiary amine results in an internal charge transfer. The colour of the emission observed is yellow. Upon addition of a strong base, the phenolic hydroxyl group is rendered deprotonated, resulting in a photoinduced electron transfer, which in turn renders the molecule non-emissive. Upon addition of both an acid and a base, the emission of the molecule is observed as red, as the tertiary amine would not be protonated while the hydroxyl group would remain protonated resulting in the absence of both PET and ICT. Due to the great difference in emission intensity, this single molecule is capable of carrying out an arithmetic operation; subtraction at a nanoscale level.

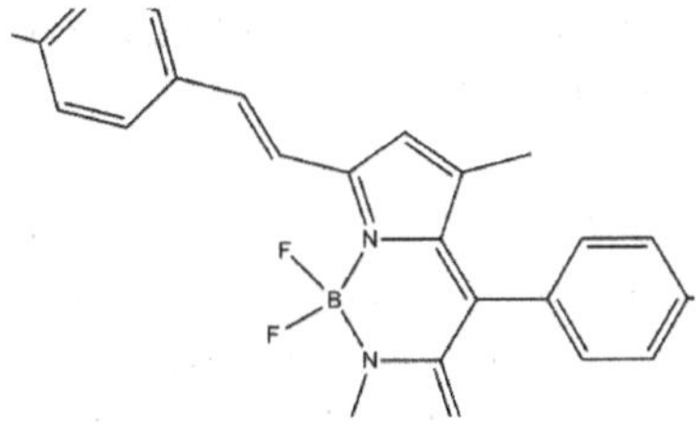

NANOCIRCUITRY

Nanocircuits are electrical circuits operating on the nanometer scale. This is well into the quantum realm, where quantum mechanical effects become very important. One nanometer is equal to 10^{-9} meters or a row of 10 hydrogen atoms. With such progressively smaller circuits, more can be fitted on a computer chip. This allows faster and more complex functions using less power. Nanocircuits are composed of three different fundamental components. These are transistors, interconnections, and architecture, all fabricated on the nanometer scale.

Various Approaches to Nanocircuitry

A variety of proposals have been made to implement nanocircuitry in different forms. These include Single-Electron Transistors, Quantum dot cellular automata, and Nanoscale Crossbar Latches. However, likely nearer-term approaches will involve incorporation of nanomaterials to improve MOSFETs. These currently form the basis of most analog and digital circuit designs, the scaling of which drives Moore's Law. A review article covering the MOSFET design and its future was published in 2004 comparing different geometries of MOSFETs under scale reduction and noted that circular cross-section vertical channel FETs are optimal for scale reduction. This configuration is capable of being implemented with a high density using vertical semiconductor cylindrical channels with nanoscale diameters and Infineon Technologies and Samsung have begun research and development in this direction resulting in some basic patents using nanowires and carbon nanotubes in MOSFET designs. In an alternative approach, Nanosys uses solution based deposition and alignment processes to pattern pre-fabricated arrays of nanowires on a substrate to serve as a lateral channel of an FET. While not capable of the same scalability as single nanowire FETs, the use of pre-fabricated multiple nanowires for the channel increases reliability and reduces production costs since large volume printing processes may be used to deposit the nanowires at a lower temperature than conventional fabrication procedures. In addition, due to the lower temperature deposition a wider variety of materials such as polymers may be used as the carrier substrate for the transistors opening the door to flexible electronic applications such as electronic paper, bendable flat panel displays, and wide area solar cells.

Production Methods

One of the most fundamental concepts to understanding nanocircuits is the formulation of

Moore's Law. This concept arose when Intel co-founder Gordon Moore became interested in the cost of transistors and trying to fit more onto one chip. It relates that the number of transistors that can be fabricated on a silicon integrated circuit—and therefore the computing abilities of such a circuit—is doubling every 18 to 24 months. The more transistors one can fit on a circuit, the more computational abilities the computer will have. This is why scientists and engineers are working together to produce these nanocircuits so millions and perhaps even billions of transistors will be able to fit onto a chip. Despite how good this may sound, there are many problems that arise when so many transistors are packed together. With circuits being so tiny, they tend to have more problems than larger circuits, more particularly heat - the amount of power applied over a smaller surface area makes heat dissipation difficult, this excess heat will cause errors and can destroy the chip. Nanoscale circuits are more sensitive to temperature changes, cosmic rays and electromagnetic interference than today's circuits. As more transistors are packed onto a chip, phenomena such as stray signals on the chip, the need to dissipate the heat from so many closely packed devices, tunneling across insulation barriers due to the small scale, and fabrication difficulties will halt or severely slow progress. Many believe the market for nanocircuits will reach equilibrium around 2015. At this time they believe the cost of a fabrication facility may be as much as $200 billion. There will be a time when the cost of making circuits even smaller will be too much, and the speed of computers will reach a maximum. For this reason, many scientists believe that Moore's Law will not hold forever and will soon reach a peak, since Moore's law is largely predicated on computational gains caused by improvements in micro-lithographic etching technologies.

In producing these nanocircuits, there are many aspects involved. The first part of their organization begins with transistors. As of right now, most electronics are using silicon-based transistors. Transistors are an integral part of circuits as they control the flow of electricity and transform weak electrical signals to strong ones. They also control electric current as they can turn it on off, or even amplify signals. Circuits now use silicon as a transistor because it can easily be switched between conducting and nonconducting states. However, in nanoelectronics, transistors might be organic molecules or nanoscale inorganic structures. Semiconductors, which are part of transistors, are also being made of organic molecules in the nano state. The second aspect of nanocircuit organization is interconnection. This involves logical and mathematical operations and the wires linking the transistors together that make this possible. In nanocircuits, nanotubes and other wires as narrow as one nanometer are used to link transistors together. Nanowires have been made from carbon nanotubes for a few years. Until a few years ago, transistors and nanowires were put together to produce the circuit. However, scientists have been able to produce a nanowire with transistors in it. In 2004, Harvard University nanotech pioneer Charles Lieber and his team have made a nanowire—10,000 times thinner than a sheet of paper—that contains a string of transistors. Essentially, transistors and nanowires are already pre-wired so as to eliminate the difficult task of trying to connect transistors together with nanowires. The last part of nanocircuit organization is architecture. This has been explained as the overall way the transistors are interconnected, so that the circuit can plug into a computer or other system and operate independently of the lower-level details. With nanocircuits being so small, they are destined for error and defects. Scientists have devised a way to get around this. Their architecture combines circuits that have redundant logic gates and interconnections with the ability to reconfigure structures at several levels on a chip. The redundancy lets the circuit identify problems and reconfigure itself so the circuit can avoid more problems. It also allows for errors within the logic gate and still have it work properly without giving a wrong result.

Potential Applications and Breakthroughs

Scientists in India have developed the world's smallest transistor which will be used for nanocircuits. The transistor is made entirely from carbon nanotubes. Nanotubes are rolled up sheets of carbon atoms and are more than a thousand times thinner than human hair. Normally circuits use silicon-based transistors, but these will soon replace those. The transistor has two different branches that meet at a single point, hence giving it a Y shape. Current can flow throughout both branches and is controlled by a third branch that turns the voltage on or off. This new breakthrough can now allow for nanocircuits to hold completely to their name as they can be made entirely from nanotubes. Before this discovery, logic circuits used nanotubes, but needed metal gates to be able to control the flow of electric current. Arguably the biggest potential application of nanocircuits deals with computers and electronics. Scientists and engineers are always looking to make computers faster. Some think in the nearer term, we could see hybrids of micro- and nano-: silicon with a nano core—perhaps a high-density computer memory that retains its contents forever. Unlike conventional circuit design, which proceeds from blueprint to photographic pattern to chip, nanocircuit design will probably begin with the chip—a haphazard jumble of as many as 1024 components and wires, not all of which will even work—and gradually sculpt it into a useful device. Instead of taking the traditional top-down approach, the bottom-up approach will probably soon have to be adopted because of the sheer size of these nanocircuits. Not everything in the circuit will probably work because at the nano level, nanocircuits will be more defective and faulty because of their compactness. Scientists and engineers have created all of the essential components of nanocircuits such as transistors, logic gates and diodes. They have all been constructed from organic molecules, carbon nanotubes and nanowire semiconductors. The only thing left to do is find a way to eliminate the errors that come with such a small device and nanocircuits will become a way of all electronics. However, eventually there will be a limit as to how small nanocircuits can become and computers and electronics will reach their equilibrium speeds.

MOLECULAR WIRES

Molecular wires (or sometimes called molecular nanowires) are molecular chains that conduct electric current. They are the proposed building blocks for molecular electronic devices. Their typical diameters are less than three nanometers, while their lengths may be macroscopic, extending to centimeters or more.

Examples

Most types of molecular wires are derived from organic molecules. One naturally occurring molecular wire is DNA. Prominent inorganic examples include polymeric materials such as $Li_2Mo_6Se_6$ and $Mo_6S_{9-x}I_x$, $[Pd_4(CO)_4(OAc)_4Pd(acac)_2]$, and single-molecule extended metal atom chains (EMACs) which comprise strings of transition metal atoms directly bonded to each other. Molecular wires containing paramagnetic inorganic moieties can exhibit Kondo peaks.

Structure

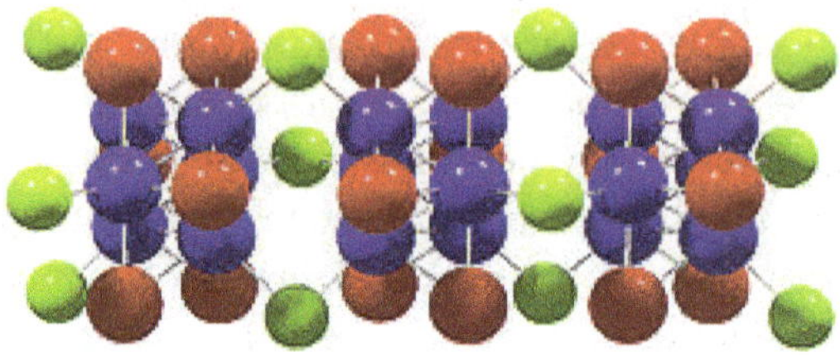

The structure of a $Mo_6S_{9-x}I_x$ molecular wire. Mo atoms are blue, iodine atoms are red and sulphur atoms are yellow.

Unlike the more usual nanowires (which are very thin crystals), molecular nanowires are composed of repeating monomers. In the case of DNA, the repeat units are the nucleotides with a backbone made of sugars and phosphate groups joined by ester bonds. Attached to each sugar is one of four types of bases. In case of $Mo_6S_{9-x}I_x$, the repeat units are $Mo_6S_{9-x}I_x$ clusters, which are joined together by flexible sulfur or iodine bridges. Molecular nanowires can be manipulated and investigated as single molecules, but they often aggregate in solution into swatches or bundles. In the case of the Mo chalcogenide-halides, they grow in the form of ordered strands, in which the individual strands are linked by very weak Van der Waals forces. Individual molecules can be manipulated, ordered and their length can be controlled with atomic force microsope tips.

Conduction of Electrons

Molecular wires conduct electricity. They typically have non-linear current-voltage characteristics, and do not behave as simple ohmic conductors. The conductance follows typical power law behavior as a function of temperature or electric field, whichever is the greater, arising from their strong one-dimensional character. Numerous theoretical ideas have been used in an attempt to understand the conductivity of one-dimensional systems, where strong interactions between electrons lead to departures from normal metallic (Fermi liquid) behavior. Important concepts are those introduced by Tomonaga, Luttinger and Wigner. Effects caused by classical Coulomb repulsion (called Coulomb blockade), interactions with vibrational degrees of freedom (called phonons) and Quantum Decoherence have also been found to be important in determining the properties of molecular wires.

Nanowires in Molecular Electronics

To be of use for connecting molecules, MWs need to self-assemble following well-defined routes and form reliable electrical contacts between them. To reproducibly self-assemble a complex circuit based on single molecules. Ideally, they would connect to diverse materials, such as gold metal surfaces (for connections to outside world), biomolecules (for nanosensors, nanoelectrodes, molecular switches) and most importantly, they must allow branching. The connectors should also be available of pre-determined diameter and length. They should also have covalent bonding to ensure reproducible transport and contact properties. DNA-like molecules have specific molecular-scale recognition and can be used in molecular scaffold fabrication. Complex shapes have been demonstrated, but unfortunately metal coated DNA which is electrically conducting is too thick to connect to individual molecules. Thinner coated DNA lacks electronic connectivity and is unsuited for connecting molecular electronics components. Some varieties of carbon nanotubes (CNTs) are conducting, and connectivity at their ends can be achieved by attachment of connecting groups. Unfortunately manufacturing CNTs with pre-determined properties is impossible at present, and the functionalized ends are typically not conducting, limiting their usefulness as molecular connectors. Individual CNTs can be soldered in an electron microscope, but the contact is not covalent and cannot be self-assembled. Possible routes for the construction of larger functional circuits using $Mo_6S_{9-x}I_x$ MWs have been demonstrated, either via gold nanoparticles as linkers, or by direct connection to thiolated molecules. The two approaches may lead to different possible applications. The use of GNPs offers the possibility of branching and construction of larger circuits.

Other Research

Molecular wires can be incorporated into polymers, enhancing their mechanical and/or conducting properties. The enhancement of these properties relies on uniform dispersion of the wires into the host polymer. MoSI wires have been made in such composites, relying on their superior solubility within the polymer host compared to other nanowires or nanotubes. Bundles of wires can be used to enhance tribological properties of polymers, with applications in actuators and potentiometers. It has been recently proposed that twisted nanowires could work as electromechanical nanodevices (or torsion nanobalances) to measure forces and torques at nanoscale with great precision.

NANOSENSORS

Nanosensors are any biological, chemical, or surgical sensory points used to convey information about nanoparticles to the macroscopic world. Their use mainly include various medicinal purposes and as gateways to building other nanoproducts, such as computer chips that work at the nanoscale and nanorobots. There are several ways being proposed today to make nanosensors; these include top-down lithography, bottom-up assembly, and molecular self-assembly.

Predicted Applications

Medicinal uses of nanosensors mainly revolve around the potential of nanosensors to accurately identify particular cells or places in the body in need. By measuring changes in volume, concentration, displacement and velocity, gravitational, electrical, and magnetic forces, pressure, or temperature of cells in a body, nanosensors may be able to distinguish between and recognize certain cells, most notably those of cancer, at the molecular level in order to deliver medicine or monitor development to specific places in the body. For example, light-responsive nanosensors can be designed to determine local protease activity in a body to detect immune response or cancer. In addition, they may be able to detect macroscopic variations from outside the body and communicate these changes to other nanoproducts working within the body.

One example of nanosensors involves using the fluorescence properties of cadmium selenide quantum dots as sensors to uncover tumors within the body. By injecting a body with these quantum dots, a doctor could see where a tumor or cancer cell was by finding the injected quantum dots, an easy process because of their fluorescence. Developed nanosensor quantum dots would be specifically constructed to find only the particular cell for which the body was at risk. A downside to the cadmium selenide dots, however, is that they are highly toxic to the body. As a result, researchers are working on developing alternate dots made out of a different, less toxic material while still retaining some of the fluorescence properties. In particular, they have been investigating the particular benefits of zinc sulfide quantum dots which, though they are not quite as fluorescent as cadmium selenide, can be augmented with other metals including manganese and various lanthanide elements. In addition, these newer quantum dots become more fluorescent when they bond to their target cells. (Quantum) Potential predicted functions may also include sensors used to detect specific DNA in order to recognize explicit genetic defects, especially for individuals at high-risk and implanted sensors that can automatically detect glucose levels for diabetic subjects more simply than current detectors. DNA can also serve as sacrificial layer for manufacturing CMOS IC, integrating a nanodevice with sensing capabilities. Therefore, using proteomic patterns and new hybrid materials, nanobiosensors can also be used to enable components configured into a hybrid semiconductor substrate as part of the circuit assembly. The development and miniaturization of nanobiosensors should provide interesting new opportunities. Other projected products most commonly involve using nanosensors to build smaller integrated circuits, as well as incorporating them into various other commodities made using other forms of nanotechnology for use in a variety of situations including transportation, communication, improvements in structural integrity, and robotics. Nanosensors may also eventually be valuable as more accurate monitors of material states for use in systems where size and weight are constrained, such as in satellites and other aeronautic machines.

Existing Examples

Currently, the most common mass-produced functioning nanosensors exist in the biological world as natural receptors of outside stimulation. For instance, sense of smell, especially in animals in which it is particularly strong, such as dogs, functions using receptors that sense nanosized molecules. Certain plants, too, use nanosensors to detect sunlight; various fish use nanosensors to detect minuscule vibrations in the surrounding water; and many insects detect sex pheromones using nanosensors. One of the first working examples of a synthetic nanosensor was built by researchers at the Georgia Institute of Technology in 1999. It involved attaching a single particle onto the end of a carbon nanotube and measuring the vibrational frequency of the nanotube both with and without the particle. The discrepancy between the two frequencies allowed the researchers to measure the mass of the attached particle. Chemical sensors, too, have been built using nanotubes to detect various properties of gaseous molecules. Carbon nanotubes have been used to sense ionization of gaseous molecules while nanotubes made out of titanium have been employed to detect atmospheric concentrations of hydrogen at the molecular level. Many of these involve a system by which nanosensors are built to have a specific pocket for another molecule. When that particular molecule, and only that specific molecule, fits into the nanosensor, and light is shone upon the nanosensor, it will reflect different wavelengths of light and, thus, be a different color. In a similar fashion, Flood et al. have shown that supramolecular host-guest chemistry offers quantitative sensing using Raman scattered light as well as SERS. Photonic devices can also be used as nanosensors to quantify concentrations of clinically relevant samples. A principle of operation of these sensors is based on the chemical modulation of a hydrogel film volume that incorporates a Bragg grating. As the hydrogel swells or shrinks upon chemical stimulation, the Bragg grating changes color and diffracts light at different wavelengths. The diffracted light can be correlated with the concentration of a target analyte.

Production Methods

There are currently several hypothesized ways to produce nanosensors. Top-down lithography is the manner in which most integrated circuits are now made. It involves starting out with a larger block of some material and carving out the desired form. These carved out devices, notably put to use in specific microelectromechanical systems used as microsensors, generally only reach the micro size, but the most recent of these have begun to incorporate nanosized components. Another way to produce nanosensors is through the bottom-up method, which involves assembling the sensors out of even more minuscule components, most likely individual atoms or molecules. This would involve moving atoms of a particular substance one by one into particular positions which, though it has been achieved in laboratory tests using tools such as atomic force microscopes, is still a significant difficulty, especially to do en masse, both for logistic reasons as well as economic ones. Most likely, this process would be used mainly for building starter molecules for self-assembling sensors.

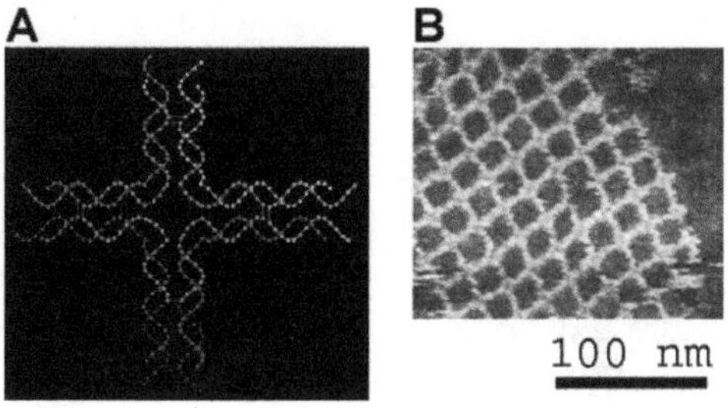

(A) An example of a DNA molecule used as a starter for larger self-assembly. (B) An atomic force microscope image of a self-assembled DNA nanogrid. Individual DNA tiles self-assemble into a highly ordered periodic two-dimensional DNA nanogrid.

The third way, which promises far faster results, involves self-assembly, or "growing" particular nanostructures to be used as sensors. This most often entails one of two types of assembly. The first involves using a piece of some previously created or naturally formed nanostructure and immersing it in free atoms of its own kind. After a given period, the structure, having an irregular surface that would make it prone to attracting more molecules as a continuation of its current pattern, would capture some of the free atoms and continue to form more of itself to make larger components of nanosensors.

The second type of self-assembly starts with an already complete set of components that would automatically assemble themselves into a finished product. Though this has been so far successful only in assembling computer chips at the micro size, researchers hope to eventually be able to do it at the nanometer size for multiple products, including nanosensors. Accurately being able to reproduce this effect for a desired sensor in a laboratory would imply that scientists could manufacture nanosensors much more quickly and potentially far more cheaply by letting numerous molecules assemble themselves with little or no outside influence, rather than having to manually assemble each sensor.

Economic Impacts

Though nanosensor technology is a relatively new field, global projections for sales of products incorporating nanosensors range from \$0.6 billion to \$2.7 billion in the next three to four years. They will likely be included in most modern circuitry used in advanced computing systems, since their potential to provide the link between other forms of nanotechnology and the macroscopic world allows developers to fully exploit the potential of nanotechnology to miniaturize computer chips while vastly expanding their storage potential.

First, however, nanosensor developers must overcome the present high costs of production in order to become worthwhile for implementation in consumer products. Additionally, nanosensor reliability is not yet suitable for widespread use, and, because of their scarcity, nanosensors have yet to be marketed and implemented outside of research facilities. Consequently, nanosensors have yet to be made compatible with most consumer technologies for which they have been projected to eventually enhance.

Social Impacts

Ethical and social impacts are harder to define and sort as good or bad compared to health and environmental impacts. The advancement in detecting and sensing different biological and chemical species with increased capacity and accuracy may transform societal mechanisms that were originally designed on uncertainty and imprecise information. For example, the ability to measure extremely low amounts of air pollutants or toxic materials in water raises questions and dilemmas of risk thresholds especially if the advancement of such technologies outpaces the ability of the public to respond. As another example, medical sensors will not only help in diagnoses and treatment but may also predict the future profile of an individual. This will add to the information used by health insurance companies to grant or deny coverage. Other social issues resulting from the widespread use of nanosensors and surveillance devices include privacy invasion and security issues.

NANOIONICS

Nanoionics is the study and application of phenomena, properties, effects and mechanisms of processes connected with fast ion transport (FIT) in all-solid-state nanoscale systems. The topics of interest include fundamental properties of oxide ceramics at nanometer length scales, and fast ion conductor (advanced superionic conductor)/electronic conductor heterostructures. Potential applications are in electrochemical devices (electrical double layer devices) for conversion and storage of energy, charge and information. The term and conception of nanoionics (as a new branch of science) were first introduced by A.L. Despotuli and V.I. Nikolaichik (Institute of Microelectronics Technology and High Purity Materials, Russian Academy of Sciences, Chernogolovka) in January 1992.

A multidisciplinary scientific and industrial field of Solid state ionics, dealing with ionic transport phenomena in solids, considers Nanoionics as its new division. Nanoionics tries to describe, for example, diffusion&reactions, in terms which make sence only at a nanoscale, e.g., in terms of non-uniform (at a nanoscale) potential landscape.

There are two classes of solid state ionic nanosystems and two fundamentally different nanoionics: (I) nanosystems based on solids with low ionic conductivity, and (II) nanosystems based on advanced superionic conductors (e.g. alpha–AgI, rubidium silver iodide–family). Nanoionics-I and nanoionics-II differ from each other in the design of interfaces. The role of boundaries in nanoionics-I is the creation of conditions for high concentrations of charged defects (vacancies and interstitials) in a disordered space-charge layer. But in nanoionics-II, it is necessary to conserve the original highly ionic conductive crystal structures of advanced superionic conductors at ordered (lattice-matched) heteroboundaries. Nanoionic-I can significantly enhance (up to $\sim10^8$ times) the 2D-like ion conductivity in nanostructured materials with structural coherence, but it is remaining $\sim10^3$ times smaller relatively to 3D ionic conductivity of advanced superionic conductors.

The classical theory of diffusion and migration in solids is based on the notion of a diffusion coefficient, activation energy and electrochemical potential. This means that accepted is the picture of a hopping ion transport in the potential landscape where all barriers are of the same height (uniform potential relief). In spite of the obvious difference of objects of Solid state ionics and nanoionics-I, -II, the true new problem of fast ion transport and charge/energy storage (or transformation) for these objects (fast ion conductors) has a special common basis: non-uniform potential landscape on nanoscale (for example) which determines the character of the mobile ion subsystem response to an impulse or harmonic external influence, e.g. a weak influence in Dielectric spectroscopy (impedance spectroscopy).

Characteristics

Being a branch of nanoscience and nanotechnology, nanoionics is unambiguously defined by its own objects (nanostructures with FIT), subject matter (properties, phenomena, effects, mechanisms of processes, and applications connected with FIT at nano-scale), method (interface design in nanosystems of superionic conductors), and criterion (R/L ~1, where R is the length scale of device structures, and L is the characteristic length on which the properties, characteristics, and other parameters connected with FIT change drastically).

The International Technology Roadmap for Semiconductors (ITRS) relates nanoionics-based resistive switching memories to the category of "emerging research devices" ("ionic memory"). The area of close intersection of nanoelectronics and nanoionics can be called nanoelionics. Now, the vision of future nanoelectronics constrained solely by fundamental ultimate limits is being formed in advanced researches. The ultimate physical limits to computation are very far beyond the currently attained (10^{10} cm^{-2}, 10^{10} Hz) region. What kind of logic switches might be used at the near nm- and sub-nm peta-scale integration? The question was the subject matter already in, where the term "nanoelectronics" was not used yet. Quantum mechanics constrains electronic distinguishable configurations by the tunneling effect at tera-scale. To overcome 10^{12} cm^{-2} bit density limit, atomic and ion configurations with acharacteristic dimension of L <2 nm should be used in the information domain and materials with an effective mass of information carriers m* considerably larger than electronic ones are required: $m^* = 13\ m_e$ at L =1 nm, $m^* = 53\ m_e$ (L =0,5 nm) and $m^* = 336\ m_e$ (L =0,2 nm). Future short-sized devices may be nanoionic, i.e. based on the fast ion transport at the nanoscale, as it was first stated in.

Examples

The examples of nanoionic devices are all-solid-state supercapacitors with fast ion transport at the functional heterojunctions (nanoionic supercapacitors), lithium batteries and fuel cells with nanostructured electrodes, nano-switches with quantized conductivity on the basis of fast ion conductors (see also memristors and programmable metallization cell). These are well compatible with sub-voltage and deep-sub-voltage nanoelectronics (see http://www.nanometer.ru/2008/02/08/nanoelektronika_5900.html) and could find wide applications, for example in autonomous micro power sources, RFID, MEMS, smartdust, nanomorphic cell, other micro- and nanosystems, or reconfigurable memory cell arrays. An important case of fast ionic conduction in solid states is that in surface space-charge layer of ionic crystals. Such conduction was first predicted by Kurt Lehovec. A significant role of boundary conditions with respect to ionic conductivity was first experimentally discovered by C.C. Liang who found an anomalously high conduction in the LiI-Al_2O_3 two-phase system. Because a space-charge layer with specific properties has nanometer thickness, the effect is directly related to nanoionics (nanoionics-I). The Lehovec effect has become the basis for the creation of a multitude of nanostructured fast ion conductors which are used in modern portable lithium batteries and fuel cells. Recently, a 1D structure-dynamic approach was developed in nanoionics for detailed description of the space charge formation and relaxation processes in irregular potential relief (direct problem) and interpretation of characteristics of nanosystems with fast ion transport (inverse problem), as example, for the description of a collective phenomenon: coupled ion transport and dielectric-polarization processes which lead to A. K. Jonscher's "universal" dynamic response.

REFERENCES

1. Petty, M.C.; Bryce, M.R. & Bloor, D. (1995). Introduction to Molecular Electronics. New York: Oxford University Press. pp. 1–25. ISBN 0-19-521156-1.

2. Tour, James M.; et al. (1998). "Recent advances in molecular scale electronics". Annals of the New York Academy of Sciences. 852: 197–204. Bibcode:1998NYASA.852..197T. doi:10.1111/j.1749-6632.1998.tb09873.x.

3. Waser, Rainer; Lüssem, B. & Bjørnholm, T. (2008). "Chapter 8: Concepts in Single-Molecule Electronics". Nanotechnology. Volume 4: Information Technology II. Weinheim: Wiley-VCH Verlag GmbH & Co. KGaA. pp. 175–212. ISBN 978-3-527-31737-0.

4. Kubatkin, S.; et al. (2003). "Single-electron transistor of a single organic molecule with access to several redox states". Nature. 425 *(6959): 698–701. Bibcode*:2003Natur.425..698K. doi:10.1038/nature02010. PMID 14562098.

5. Anderson, Mark (2005-06-09) "Honey, I Shrunk the PC". Wired.com

6. Zoldan, Vinícius Claudio; Faccio, Ricardo & Pasa, André Avelino (2015). "N and p type character of single molecule diodes". Scientific Reports. 5: 8350. Bibcode:2015NatSR...5E8350Z. doi:10.1038/srep08350. PMC 4322354. PMID 25666850.

7. Gimzewski, J.K.; Joachim, C. (1999). "Nanoscale science of single molecules using local probes". Science. 283 *(5408): 1683–1688. Bibcode*:1999Sci...283.1683G. doi:10.1126/science.283.5408.1683. PMID 10073926.

8. Sørensen, J.K.. (2006). "Synthesis of new components, functionalized with (60) fullerene, for molecular electronics". 4th Annual meeting – CONT 2006, University of Copenhagen.

9. Service, R.F. (2003). "Molecular electronics – Next-generation technology hits an early midlife crisis". Science. 302 *(5645): 556–+. doi:10.1126/science.302.5645.556. PMID 14576398.*

10. Aviram, Arieh; Ratner, Mark A. (1974). "Molecular rectifiers". Chemical Physics Letters. 29 *(2): 277–283. Bibcode*:1974CPL....29..277A. doi:10.1016/0009-2614(74)85031-1.

11. Carter, F. L.; Siatkowski, R. E. and Wohltjen, H. (eds.) (1988) Molecular Electronic Devices, pp. 229–244, North Holland, Amsterdam.

12. Gupta, Chaitanya; Shannon, Mark A.; Kenis, Paul J. A. (2009). "Mechanisms of Charge Transport through Monolayer-Modified Polycrystalline Gold Electrodes in the Absence of Redox-Active Moieties". The Journal of Physical Chemistry C. 113 *(11): 4687–4705.* doi:10.1021/jp8090045.

13. Gupta, Chaitanya; Shannon, Mark A.; Kenis, Paul J. A. (2009). "Electronic Properties of a Monolayer–Electrolyte Interface Obtained from Mechanistic Impedance Analysis". The Journal of Physical Chemistry C. 113 *(21): 9375–9391.* doi:10.1021/jp900918u.

14. Aviram, Arieh; Ratner, M.A. (1974). "Molecular rectifiers". Chemical Physics Letters. 29 *(2): 277–283. Bibcode*:1974CPL....29..277A. doi:10.1016/0009-2614(74)85031-1.

15. Reed, M.A.; et al. (1997). "Conductance of a Molecular Junction". Science. 287 *(5336): 252–254. doi:10.1126/science.278.5336.252.*

16. Jacoby, Mitch (27 January 2003). "Molecule-based circuitry revisited". Chemical and Engineering News. Retrieved 24 February 2011.

17. Ashwell, Geoffrey J.; Hamilton, Richard; High, L. R. Hermann (2003). "Molecular rectification: asymmetric current-voltage curves from self-assembled monolayers of a donor-(n-bridge)-acceptor dye". Journal of Materials Chemistry. 13 *(7): 1501.* doi:10.1039/B304465N.

18. Ashwell, Geoffrey J.; Chwialkowska, Anna; High, L. R. Hermann (2004). "Au-S-C_nH_{2n}-Q3CNQ: self-assembled monolayers for molecular rectification". Journal of Materials Chemistry. 14 *(15): 2389.* doi:10.1039/B403942D.

19. Ashwell, Geoffrey J.; Chwialkowska, Anna; Hermann High, L. R. (2004). "Rectifying Au-S-C_nH_{2n}-P3CNQ derivatives". Journal of Materials Chemistry. 14 *(19): 2848.* doi:10.1039/B411343H.

20. Bumm, L. A.; Arnold, J. J.; Cygan, M. T.; Dunbar, T. D.; Burgin, T. P.; Jones, L.; Allara, D. L.; Tour, J. M.; Weiss, P. S. (1996). "Are Single Molecular Wires Conducting?". *Science.* 271 *(5256): 1705–1707. Bibcode*:1996Sci...271.1705B. doi:10.1126/science.271.5256.1705.

CHAPTER 4

Nanobiotechnology

Chapter Contents

- **Introduction**
- **Applications**
- **Tools**

Nanotechnology is defined as the study and use of structures between 1 nanometer and 100 nanometers in size. To give you an idea of how small that is, it would take eight hundred 100 nanometer particles side by side to match the width of a human hair.

INTRODUCTION

Nanobiotechnology, bionanotechnology, and nanobiology are terms that refer to the intersection of nanotechnology and biology. Given that the subject is one that has only emerged very recently, bionanotechnology and nanobiotechnology serve as blanket terms for various related technologies.

This discipline helps to indicate the merger of biological research with various fields of nanotechnology. Concepts that are enhanced through nanobiology include: nanodevices (such as biological machines), nanoparticles, and nanoscale phenomena that occurs within the discipline of nanotechnology. This technical approach to biology allows scientists to imagine and create systems that can be used for biological research. Biologically inspired nanotechnology uses biological systems as the inspirations for technologies not yet created.However, as with nanotechnology and biotechnology, bionanotechnology does have many potential ethical issues associated with it.

The most important objectives that are frequently found in nanobiology involve applying nanotools to relevant medical/biological problems and refining these applications. Developing new tools, such as peptoid nanosheets, for medical and biological purposes is another primary objective in nanotechnology. New nanotools are often made by refining the applications of the nanotools that are already being used. The imaging of native biomolecules, biological membranes, and tissues is also a major topic for the nanobiology researchers. Other topics concerning nanobiology include the use of cantilever array sensors and the application of nanophotonics for manipulating molecular processes in living cells. Recently, the use of microorganisms to synthesize functional nanoparticles has been of great interest. Microorganisms can change the oxidation state of metals. These microbial processes have opened up new opportunities for us to explore novel applications, for example, the biosynthesis of metal nanomaterials. In contrast to chemical and physical methods, microbial processes for synthesizing nanomaterials can be achieved in aqueous phase under gentle and environmentally benign conditions. This approach has become an attractive focus in current green bionanotechnology research towards sustainable development.

Terminology

The terms are often used interchangeably. When a distinction is intended, though, it is based on whether the focus is on applying biological ideas or on studying biology with nanotechnology. Bionanotechnology generally refers to the study of how the goals of nanotechnology can be guided by studying how biological "machines" work and adapting these biological motifs into improving existing nanotechnologies or creating new ones. Nanobiotechnology, on the other hand, refers to the ways that nanotechnology is used to create devices to study biological systems. In other words, nanobiotechnology is essentially miniaturized biotechnology, whereas bionanotechnology is a specific application of nanotechnology. For example, DNA nanotechnology or cellular engineering would be classified as bionanotechnology because they involve working with biomolecules on the nanoscale. Conversely, many new medical technologies involving nanoparticles as delivery systems or as sensors would be examples of nanobiotechnology since they involve using nanotechnology to advance the goals of biology. The definitions enumerated above will be utilized whenever a distinction between nanobio and bionano is made in this article. However, given the overlapping usage of the terms in modern parlance, individual technologies may need to be evaluated to determine which term is more fitting. As such, they are best discussed in parallel.

Concepts

Most of the scientific concepts in bionanotechnology are derived from other fields. Biochemical principles that are used to understand the material properties of biological systems are central in bionanotechnology because those same principles are to be used to create new technologies. Material properties and applications studied in bionanoscience include mechanical properties(e.g. deformation, adhesion, failure), electrical/electronic (e.g. electromechanical stimulation, capacitors, energy storage/batteries), optical (e.g.

absorption, luminescence, photochemistry), thermal (e.g. thermomutability, thermal management), biological (e.g. how cells interact with nanomaterials, molecular flaws/defects, biosensing, biological mechanisms s.a. mechanosensing), nanoscience of disease (e.g. genetic disease, cancer, organ/tissue failure), as well as computing (e.g. DNA computing) and agriculture(target delivery of pesticides, hormones and fertilizers. The impact of bionanoscience, achieved through structural and mechanistic analyses of biological processes at nanoscale, is their translation into synthetic and technological applications through nanotechnology. Nano-biotechnology takes most of its fundamentals from nanotechnology. Most of the devices designed for nano-biotechnological use are directly based on other existing nanotechnologies. Nano-biotechnology is often used to describe the overlapping multidisciplinary activities associated with biosensors, particularly where photonics, chemistry, biology, biophysics, nano-medicine, and engineering converge. Measurement in biology using wave guide techniques, such as dual polarization interferometry, are another example.

APPLICATIONS

Applications of bionanotechnology are extremely widespread. Insofar as the distinction holds, nanobiotechnology is much more commonplace in that it simply provides more tools for the study of biology. Bionanotechnology, on the other hand, promises to recreate biological mechanisms and pathways in a form that is useful in other ways.

Nanomedicine

Nanomedicine is a field of medical science whose applications are increasing more and more thanks to nanorobots and biological machines, which constitute a very useful tool to develop this area of knowledge. In the past years, researchers have done many improvements in the different devices and systems required to develop nanorobots. This supposes a new way of treating and dealing with diseases such as cancer; thanks to nanorobots, side effects of chemotherapy have been controlled, reduced and even eliminated, so some years from now, cancer patients will be offered an alternative to treat this disease instead of chemotherapy, which causes secondary effects such as hair loss, fatigue or nausea killing not only cancerous cells but also the healthy ones. At a clinical level, cancer treatment with nanomedicine will consist on the supply of nanorobots to the patient through an injection that will seek for cancerous cells leaving untouched the healthy ones. Patients that will be treated through nanomedicine will not notice the presence of this nanomachines inside them; the only thing that is going to be noticeable is the progressive improvement of their health.

Nanobiotechnology

Nanobiotechnology (sometimes referred to as nanobiology) is best described as helping modern medicine progress from treating symptoms to generating cures and regenerating biological tissues. Three American patients have received whole cultured bladders with the help of doctors who use nanobiology techniques in their practice. Also, it has been demonstrated in animal studies that a uterus can be grown outside the body and then placed in the body in order to produce a baby. Stem cell treatments have been used to fix diseases that are found in the human heart and are in clinical trials in the United States. There is also funding for research into allowing people to have new limbs without having to resort to prosthesis. Artificial proteins might also become available to manufacture without the need for harsh chemicals and expensive machines. It has even been surmised that by the year 2055, computers may be made out of biochemicals and organic salts.

Another example of current nanobiotechnological research involves nanospheres coated with fluorescent polymers. Researchers are seeking to design polymers whose fluorescence is quenched when they encounter specific molecules. Different polymers would detect different metabolites. The polymer-coated spheres could become part of new biological assays, and the technology might someday lead to particles which could be introduced into the human body to track down metabolites associated with tumors and other health problems. Another example, from a different perspective, would be evaluation and therapy at the nanoscopic level, i.e. the treatment of Nanobacteria (25-200 nm sized) as is done by NanoBiotech Pharma.

While nanobiology is in its infancy, there are a lot of promising methods that will rely on nanobiology in the future. Biological systems are inherently nano in scale; nanoscience must merge with biology in order to deliver biomacromolecules and molecular machines that are similar to nature. Controlling and mimicking the devices and processes that are constructed from molecules is a tremendous challenge to face the converging disciplines of nanotechnology. All living things, including humans, can be considered to be nanofoundries. Natural evolution has optimized the "natural" form of nanobiology over millions of years. In the 21st century, humans have developed the technology to artificially tap into nanobiology. This process is best described as "organic merging with synthetic." Colonies of live neurons can live together on a biochip device; according to research from Dr. Gunther Gross at the University of North Texas. Self-assembling nanotubes have the ability to be used as a structural system. They would be composed together with rhodopsins; which would facilitate the optical computing process and help with the storage of biological materials. DNA (as the software for all living things) can be used as a structural proteomic system - a logical component for molecular computing. Ned Seeman - a researcher at New York University - along with other researchers are currently researching concepts that are similar to each other.

Bionanotechnology

DNA nanotechnology is one important example of bionanotechnology. The utilization of the inherent properties of nucleic acids like DNA to create useful materials is a promising area of modern research. Another important area of research involves taking advantage of membrane properties to generate synthetic membranes. Proteins that self-assemble to generate functional materials could be used as a novel approach for the large-scale production of programmable nanomaterials. One example is the development of amyloids found in bacterial biofilms as engineered nanomaterials that can be programmed genetically to have different properties. Protein folding studies provide a third important avenue of research, but one that has been largely inhibited by our inability to predict protein folding with a sufficiently high degree of accuracy. Given the myriad uses that biological systems have for proteins, though, research into understanding protein folding is of high importance and could prove fruitful for bionanotechnology in the future.

Lipid nanotechnology is another major area of research in bionanotechnology, where physico-chemical properties of lipids such as their antifouling and self-assembly is exploited to build nanodevices with applications in medicine and engineering.

Nanobiotechnology Applications in Agriculture

Meanwhile, nanotechnology application to biotechnology will also leave no field untouched by its groundbreaking scientific innovations for human wellness; the agricultural industry is no exception. Basically, nanomaterials are distinguished depending on the origin: natural, incidental and engineered nanoparticles. Among these, engineered nanoparticles have received wide attention in all fields of science, including medical, materials and agriculture technology with significant socio-economical growth. In the agriculture industry, engineered nanoparticles have been serving as nano carrier, containing herbicides, chemicals, or genes, which target particular plant parts to release their content. Previously nanocapsules containing herbicides have been reported to effectively penetrate through cuticles and tissues, allowing the slow and constant release of the active substances. Likewise, other literature describes that nano-encapsulated slow release of fertilizers has also become a trend to save fertilizer consumption and to minimize environmental pollution through precision farming. These are only a few examples from numerous research works which might open up exciting opportunities for nanobiotechnology application in agriculture. Also, application of this kind of engineered nanoparticles to plants should be considered the level of amicability before it is employed in agriculture practices. Based on a thorough literature survey, it was understood that there is only limited authentic information available to explain the biological consequence of engineered nanoparticles on treated plants. Certain reports underline the phytotoxicity of various origin of engineered nanoparticles to the plant caused by the subject of concentrations and sizes . At the same time, however, an equal number of studies were reported with a positive outcome of nanoparticles, which facilitate growth promoting nature to treat plant. In particular, compared to other nanoparticles, silver and gold nanoparticles based applications elicited beneficial results on various plant species with less and/or no toxicity.Silver nanoparticles (AgNPs) treated leaves of Asparagus showed the increased content of ascorbate and chlorophyll. Similarly, AgNPs-treated common bean and corn has increased shoot and root length, leaf surface area, chlorophyll, carbohydrate and protein contents reported earlier. The gold nanoparticle has been used to induce growth and seed yield in Brassica juncea.

TOOLS

This field relies on a variety of research methods, including experimental tools (e.g. imaging, characterization via AFM/optical tweezers etc.), x-ray diffraction based tools, synthesis via self-assembly, characterization of self-assembly (using e.g. MP-SPR, DPI, recombinant

DNA methods, etc.), theory (e.g. statistical mechanics, nanomechanics, etc.), as well as computational approaches (bottom-up multi-scale simulation, supercomputing).

REFERENCES

1. Ehud Gazit, Plenty of room for biology at the bottom: An introduction to bionanotechnology. Imperial College Press, 2007, ISBN 978-1-86094-677-6
2. «Nanobiology". Nanotech-Now.com.
3. «Nanobiology". Swiss Nanoscience Institute.
4. Ng, CK; Sivakumar K; Liu X; Madhaiyan M; Ji L; Yang L; Tang C; Song H; Kjelleberg S; Cao B. (4 Feb 2013). "Influence of outer membrane c-type cytochromes on particle size and activity of extracellular nanoparticles produced by Shewanella oneidensis.". *Biotechnology and Bioengineering.* **110** *(7): 1831–7. doi:10.1002/bit.24856. PMID 23381725.*
5. Bionanotechnology - Definition, wordiQ.com
6. Nolting B, "Biophysical Nanotechnology". In: "Methods in Modern Biophysics", Springer, 2005, ISBN 3-540-27703-X
7. NBTC Homepage | Nanobiotechnology Center
8. Raja muthuramalingam thangavelu et al (2016) Nanobiotechnological approach using plant rooting hormones synthesized silver nanoparticle as a "nanobullets" for the dynamic applications in horticulture -An in vitro and ex vitro study, Arabian Journal of chemistry,http://dx.doi.org/10.1016/j.arabjc.2016.09.022
9. Venkatesan M, Jolad B, editors. Emerging Trends in Robotics and Communication Technologies (INTERACT). 2010 International Conference on . Nanorobots in cancer treatment; 12/3-5; Chennai: IEEE; 2010. doi:10.1109/INTERACT.2010.5706154
10. "The future of nano-biology". ZD Net.
11. "Nanobiology: from physics and engineering to biology". IOP Science.
12. "The Nanobiology Imperative". HistorianoftheFuture.com.
13. Zadegan, Reza M.; Norton, Michael L. (June 2012). "Structural DNA Nanotechnology: From Design to Applications". Int. J. Mol. Sci. **13** *(6): 7149–7162.* doi:10.3390/ijms13067149. PMC 3397516. PMID 22837684.
14. Nguyen, Peter; Botyanszki, Zsofia; Tay, Pei-Kun; Joshi, Neel (Sep 17, 2014). "Programmable biofilm-based materials from engineered curli nanofibres". Nature Communications. **5**: 4945. doi:10.1038/ncomms5945. PMID 25229329.
15. Mashaghi S.; Jadidi T.; Koenderink G.; Mashaghi A. (2013). "Lipid Nanotechnology". Int. J. Mol. Sci. **2013** *(14): 4242–4282.* doi:10.3390/ijms14024242. PMC 3588097. PMID 23429269.
16. Raja muthuramalingam thangavelu et al (2016) Nanobiotechnological approach using plant rooting hormones synthesized silver nanoparticle as a nanobullets for the dynamic applications in horticulture -An in vitro and ex vitro study, Arabian Journal of chemistry,http://dx.doi.org/10.1016/j.arabjc.2016.09.022
17. Raja muthuramalingam thangavelu et al (2016) Nanobiotechnological approach using

plant rooting hormones synthesized silver nanoparticle as a "nanobullets" for the dynamic applications in horticulture -An in vitro and ex vitro study, Arabian Journal of chemistry,http://dx.doi.org/10.1016/j.arabjc.2016.09.022

18. Raja muthuramalingam, T., Chandrasekar, S., Dharanivasan, G., Nallusamy, D., Rajendran, N., Kathiravan, K., 2015. Bioactive bile salt capped silver nanoparticle activity against destructive plant pathogenic fungi through in vitro system. RSC advances. 5, 71174-71182

19. Raqual, B., Eudald, C., Joan, C., Xavier, F., Antoni, S., Victor, P., 2009. Evaluation of the ecotoxicity of model nanoparticles. Chemosphere. 75, 850-857

20. Hediat Salama, M. H., 2012. Effects of silver nanoparticles in some crop plants, common bean (Phaseolus vulgaris L.) and corn (Zea mays L.). International research journal of biotechnology. 3(10), 190-197(

21. Sandeep arora., Priyadarshini sharma., Sumit kumar., Rajeev nayan., Khanna, P. K., Zaidi. M. G. N., 2012. Gold nanoparticles induced enhancement in growth and seed yield of Brassica juncea. Plant Growth Regul. 66, 303-310

CHAPTER 5

Nanolithography

Chapter Contents

- Introduction
- Nanoimprint Lithography
- X-Ray Lithography
- Other Techniques
- Photolithography
- Ion Beam Lithography

Nanolithography is the art and science of etching, writing, or printing at the microscopic level, where the dimensions of characters are on the order of nanometers (units of 10 $^{-9}$meter, or millionths of a millimeter). This includes various methods of modifying semiconductor chips at the atom ic level for the purpose of fabricating integrated circuits (IC s).

Instruments used in nanolithography include the scanning probe microscope (SPM) and the atomic force microscope (ATM). The SPM allows surface viewing in fine detail without necessarily modifying it. Either the SPM or the ATM can be used to etch, write, or print on a surface in single-atom dimensions.

INTRODUCTION

Nanolithography is the branch of nanotechnology concerned with the study and application of fabricating nanometer-scale structures, meaning patterns with at least one lateral dimension between 1 and 100 nm. Different approaches can be categorized in serial or parallel, mask or maskless/direct-write, top-down or bottom-up, beam or tip-based, resist-based or resist-less methods. As of 2015, nanolithography is a very active area of research in academia and in industry. Applications of nanolithography include among others: Multigate devices such as Field effect transistors (FET), Quantum dots, Nanowires, Gratings, Zone plates and Photomasks, nanoelectromechanical systems (NEMS), or semiconductorintegrated circuits (nanocircuitry).

Optical Lithography

Optical lithography, which has been the predominant patterning technique since the advent of the semiconductor age, is capable of producing sub-100-nm patterns with the use of very short optical wavelengths. Several optical lithography techniques require the use of liquid immersion and a host of resolution enhancement technologies like phase-shift masks (PSM) and optical proximity correction (OPC). Multiple patterning is a method of increasing the resolution by printing features in between pre-printed features on the same layer by etching or creating sidewall spacers, and has been used in commercial production of microprocessors since the 32 nm process node e.g. by directed self-assembly (DSA). Extreme ultraviolet lithography (EUVL) uses ultrashort wavelengths (13.5 nm) and as of 2015, is the most popularly considered Next-generation lithography (NGL) technique for mass-fabrication.

Electron-beam Lithography

Electron beam lithography or *Electron-Beam Direct-Write Lithography* (EBDW) scans a focused beam of electrons on a surface covered with an electron-sensitive film or resist (e.g. PMMA or HSQ) to draw custom shapes. By changing the solubility of the resist and subsequent selective removal of material by immersion in a solvent, sub-10 nm resolutions have been achieved. This form of direct-write, maskless lithography has high resolution and low throughput, limiting single-column e-beams to photomask fabrication, low-volume production of semiconductor devices, and research&development. Multiple-electron beam approaches have as a goal an increase of throughput for semiconductor mass-production.

NANOIMPRINT LITHOGRAPHY

Scanning probe Lithography

Scanning probe lithography (SPL) describes a set of nanolithographic methods to pattern material on the nanoscale using scanning probes. It is a direct-write, mask-less approach which bypasses the diffraction limit and can reach resolutions below 10 nm. It is considered an alternative lithographic technology often used in academic and research environments. The term *scanning probe lithography* was coined after the first patterning experiments with scanning probe microscopes (SPM) in the late 1980s.

Classification

The different approaches towards SPL can be classified by their goal to either add or remove material, by the general nature of the process either chemical or physical, or according to the driving mechanisms of the probe-surface interaction used in the patterning process: mechanical, thermal, diffusive and electrical.

Overview

Mechanical/thermo-Mechanical

Mechanical scanning probe lithography (m-SPL) is a nanomachining or *nano-scratching* top-down approach without the application of heat. **Thermo-mechanical SPL** applies heat together with a mechanical force, e.g. indenting of polymers in the Millipede memory.

Thermal

Thermal scanning probe lithography (t-SPL) uses a heatable scanning probe in order to efficiently remove material from a surface without the application of significant mechanical forces. The patterning depth can be controlled to create high-resolution 3D structures.

Thermo-Chemical

Thermochemical scanning probe lithography (tc-SPL) or *thermochemical nanolithography* (TCNL) employs the scanning probe tips to induce thermally activated chemical reactions to change the chemical functionality or the phase of surfaces. Such thermally activated reactions have been shown in proteins, organic semiconductors, electroluminescent conjugated polymers, and nanoribbon resistors. Furthermore, deprotection of functional groups (sometimes involving a temperature gradients), reduction of oxides, and the crystallization of piezoelectric/ferroelectric ceramics has been demonstrated.

Dip-pen/thermal dip-pen

Dip-pen scanning probe lithography (dp-SPL) or *dip-pen nanolithography* (DPN) is a scanning probe lithography technique based on diffusion, where the tip is employed to create patterns on a range of substances by deposition of a variety of liquid inks. **Thermal dip-pen scanning probe lithography** or *thermal dip-pen nanolithography* (TDPN) extends the usable inks to solids, which can be deposited in their liquid form when the probes are pre-heated.

Oxidation

Oxidation scanning probe lithography (o-SPL), also called *local oxidation nanolithography* (LON), *scanning probe oxidation, nano-oxidation, local anodic oxidation, AFM oxidation lithography* is based on the spatial confinement of an oxidation reaction.

Bias induced

Bias-induced scanning probe lithography (b-SPL) uses the high electrical fields created at the apex of a probe tip when voltages are applied between tip and sample to facilitate and confining a variety of chemical reactions to decompose gasesor liquids in order to locally deposit and grow materials on surfaces.

Current induced

In **Current induced scanning probe lithography** (c-SPL) in addition to the high electrical fields of b-SPL, also a focused electron current which emanates from the SPM tip is used to create nanopatterns, e.g. in polymers and molecular glasses.

Comparison to other Lithographic Techniques

Being a serial technology, SPL is inherently slower than e.g. photolithography or nanoimprint lithography, while parallelization as required for mass-fabrication is considered a large systems engineering effort (*see also Millipede memory*). As for resolution, SPL methods bypass the optical diffraction limit due to their use of scanning probes compared with photolithographic methods. Some probes have integrated in-situ metrology capabilities, allowing for feedback control during the write process. SPL works under ambient atmospheric conditions, without the need for ultra high vacuum (UHV), unlike e-beam or EUV lithography.

Multiphoton Lithography

Multiphoton lithography (also known as direct laser lithography or direct laser writing) of polymer templates has been known for years by the photonic crystal community. Similar to standard photolithography techniques, structuring is accomplished by illuminating negative-tone or positive-tone photoresists via light of a well-defined wavelength. The fundamental difference is, however, the avoidance of reticles. Instead, two-photon absorption is utilized to induce a dramatic change in the solubility of the resist for appropriate developers.

Hence, multiphoton lithography is a technique for creating small features in a photosensitive material, without the use of complex optical systems or photomasks. This method relies on a multi-photon absorption process in a material that is transparent at the wavelength of the laser used for creating the pattern. By scanning and properly modulating the laser, a chemical change (usually polymerization) occurs at the focal spot of the laser and can be controlled to create an arbitrary three-dimensional periodic or non-periodic pattern. This method has been used for rapid prototyping of structures with fine features.

Two-photon absorption is a third-order with respect to the third-order optical susceptibility $\chi^{(3)}$ and a second-order process with respect to light intensity. For this reason it is a non-linear process several orders of magnitude weaker than linear absorption, thus very high light intensities are required to increase the number of such rare events. For example, tightly-focused laser beams provide the needed intensities. Here, pulsed laser sources are preferred as they deliver high-intensity pulses while depositing a relatively low average energy. To enable 3D structuring, the light source must be adequately adapted to the photoresist in that single-photon absorption is highly suppressed while two-photon absorption is favoured. This condition is met if and only if the resist is highly transparent for the laser light's output wavelength λ and, simultaneously, absorbing at $\lambda/2$. As a result, a given sample relative to the focused laser beam can be scanned while changing the resist's solubility only in a confined volume. The geometry of the latter mainly depends on the iso-intensity surfaces of the focus. Concretely, those regions of the laser beam which exceed a given exposure threshold of the photosensitive medium define the basic building block, the so-called *voxel*. Other parameters which influence the actual shape of the voxel are the laser mode and the refractive-index mismatch between the resist and the immersion system leading to spherical aberration.

It was found that polarization effects in laser 3D nanolithography can be employed to fine-tune the feature sizes (and corresponding aspect ratio) in the structuring of photoresists. This proves polarization to be a variable parameter next to laser power (intensity), scanning speed (exposure duration), accumulated dose, etc

X-RAY LITHOGRAPHY

X-ray lithography, is a process used in electronic industry to selectively remove parts of a thin film. It uses X-rays to transfer a geometric pattern from a mask to a light-sensitive chemical photoresist, or simply "resist," on the substrate. A series of chemical treatments then engraves the produced pattern into the material underneath the photoresist.

Mechanisms

X-ray lithography originated as a candidate for next-generation lithography for the semiconductor industry, with batches of microprocessors successfully produced. Having short wavelengths (below 1 nm), X-rays overcome the diffraction limits of optical lithography, allowing smaller feature sizes. If the X-ray source isn't collimated, as with a synchrotron radiation, elementary collimating mirrors or diffractive lenses are used in the place of the refractive lenses used in optics. The X-rays illuminate a mask placed in proximity of a resist-coated wafer. The X-rays are broadband, typically from a compact synchrotron radiation source, allowing rapid exposure. Deep X-ray lithography (DXRL) uses yet shorter wavelengths on the order of 0.1 nm and modified procedures such as the LIGA process, to fabricate deep and even three-dimensional structures.

The mask consists of an X-ray absorber, typically of gold or compounds of tantalum or tungsten, on a membrane that is transparent to X-rays, typically of silicon carbide or diamond. The pattern on the mask is written by direct-write electron beam lithography onto a resist that is developed by conventional semiconductor processes. The membrane can be stretched for overlay accuracy.

Most X-ray lithography demonstrations have been performed by copying with image fidelity (without magnification) on the line of fuzzy contrast as illustrated in the figure. However, with the increasing need for high resolution, X-ray lithography is now performed on what is called the "sweet spot", using local "demagnification by bias". Dense structures are developed by multiple exposures with translation. The advantages of using

3x demagnification include, the mask is more easily fabricated, the mask to wafer gap is increased, and the contrast is higher. The technique is extensible to dense 15 nm prints.

X-rays generate secondary electrons as in the cases of extreme ultraviolet lithography and electron beam lithography. While the fine pattern definition is due principally to secondaries from Auger electrons with a short path length, the primary electrons will sensitize the resist over a larger region than the X-ray exposure. While this does not affect the pattern pitch resolution, which is determined by wavelength and gap, the image exposure contrast *(max-min)/(max+min)* is reduced because the pitch is on the order of the primary photo-electron range. The sidewall roughness and slopes are influenced by these secondary electrons as they can travel few micrometers in the area under the absorber, depending on exposure X-ray energy. Several prints at about 30 nm have been published.

Another manifestation of the photoelectron effect is exposure to X-ray generated electrons from thick gold films used for making daughter masks. Simulations suggest that photoelectron generation from the gold substrate may affect dissolution rates.

Photoelectrons, Secondary Electrons and Auger Electrons

Secondary electrons have energies of 25 eV or less, and can be generated by any ionizing radiation (VUV, EUV, X-rays, ions and other electrons). Auger electrons have energies of hundreds of electronvolts. The secondaries (generated by and outnumbering the Auger and primary photoelectrons) are the main agents for resist exposure.

The relative ranges of photoelectron primaries and Auger electrons depend on their respective energies. These energies depend on the energy of incident radiation and on the composition of the resist. There is considerable room for optimum selection (reference 3 of the article). When Auger electrons have lower energies than primary photoelectrons, they have shorter ranges. Both decay to secondaries which interact with chemical bonds. When secondary energies are too low, they fail to break the chemical bonds and cease to affect print resolution. Experiments prove that the combined range is less than 20 nm. On the other hand, the secondaries follow a different trend below ≈30 eV: the lower the energy, the longer the mean free path though they are not then able to affect resist development.

As they decay, primary photo-electrons and Auger electrons eventually become physically indistinguishable (as in Fermi–Dirac statistics) from secondary electrons. The range of low-energy secondary electrons is sometimes larger than the range of primary photo-electrons or of Auger electrons. What matters for X-ray lithography is the effective range of electrons that have sufficient energy to make or break chemical bonds in negative or positive resists.

Lithographic Electron Range

X-rays do not charge. The relatively large mean free path (~20 nm) of secondary electrons hinders resolution control at nanometer scale. In particular, electron beam lithography suffers negative charging by incident electrons and consequent beam spread which limits resolution. It is difficult therefore to isolate the effective range of secondaries which may be less than 1 nm.

The combined electron mean free path results in an image blur, which is usually modeled as a Gaussian function (where σ = blur) that is convolved with the expected image. As the desired resolution approaches the blur, the *dose image* becomes broader than the *aerial image* of the incident X-rays. The blur that matters is the *latent image* that describes the making or breaking of bonds during the exposure of resist. The *developed image* is the final relief image produced by the selected high contrast development process on the latent image.

The range of primary, Auger, secondary and ultralow energy higher-order generation electrons which print (as STM studies proved) can be large (tens of nm) or small (nm), according to various cited publications. Because this range is not a fixed number, it is hard to quantify. Line edge roughness is aggravated by the associated uncertainty. Line edge roughness is supposedly statistical in origin and only indirectly dependent on mean range. Under commonly practiced lithography conditions, the various electron ranges can be controlled and utilized.

Charging

X-rays carry no charge, but at the energies involved, Auger decay of ionized species in a specimen is more probable than radiative decay. High-energy radiation exceeding the ionization potential also generates free electrons which are negligible compared to those produced by electron beams which are charged. Charging of the sample following ionization is an extremely weak possibility when it cannot be guaranteed the ionized electrons leaving the surface or remaining in the sample are adequately balanced from other sources in time. The energy transfer to electrons as a result of ionizing radiation results in separated positive and negative charges which quickly recombine due partly to the long range of the Coulomb force. Insulating films like gate oxides and resists have been observed to charge to a positive or negative potential under electron-beam irradiation. Insulating films are eventually neutralized locally by space charge (electrons entering and exiting the surface) at the resist-vacuum interface and Fowler-Nordheim injection from the substrate. The range of the electrons in the film can be affected by the local electric field. The situation is complicated by the presence of holes (positively charged electron vacancies) which are generated along with the secondary electrons, and which may be expected to follow them around. As neutralization proceeds, any initial charge concentration effectively starts to spread out. The final chemical state of the film is reached after neutralization is completed, after all the electrons have eventually slowed down. Usually, excepting X-ray steppers, charging can be further controlled by flood gun or resist thickness or charge dissipation layer.

OTHER TECHNIQUES

- **Molecular self-assembly** (as bottom-up approach) of dense lines less than 20 nm wide in large pre-patterned trenches has been demonstrated. The degree of dimension and orientation control as well as prevention of lamella merging still need to be addressed for this to be an effective patterning technique. The important issue of line edge roughness is also highlighted by this technique. Self-assembled ripple patterns and dot arrays formed by low-energy ion-beam sputtering are another emerging form of bottom-up lithography. Aligned arrays of plasmonic and magnetic wires and nanoparticles are deposited on these templates via oblique evaporation. The templates are easily produced over large areas with periods down to 25 nm.

- **Stencil lithography** is a resist-less and parallel method of fabricating nanometer scale patterns using nanometer-size apertures as shadow-masks.
- **X-ray lithography** can be extended to a resolution of 15 nm by using the X-ray wavelengths of 1 nm as illumination. This is implemented by the proximity printing approach. The technique is developed to the extent of batch processing. The extension of the method relies on Near Field X-rays in Fresnel diffraction: A clear mask feature is "demagnified" by proximity to a wafer that is set near to a "Critical Condition". This Condition determines the mask-to-wafer gap and depends on both the size of the clear mask feature and on the wavelength. The method is simple because it requires no lenses.
- **Laser printing of single nanoparticles** In this method, the optical forces induced via scattering and absorption of photons on nanoparticles are used to direct single nanoparticles to specific locations on substrates and attach them via van-der Waals forces. This technique has been demonstrated on metallic nanoparticles, which are easier to print due to their large plasmonically-induced scattering and absorption cross sections, in both serial and parallel printing methods.
- **Magnetolithography** (ML) based on applying a magnetic field on the substrate using paramagnetic metal masks call "magnetic mask". Magnetic mask which is analog to photomask define the spatial distribution and shape of the applied magnetic field. The second component is ferromagnetic nanoparticles (analog to the Photoresist) that are assembled onto the substrate according to the field induced by the magnetic mask.
- **Nanosphere lithography** uses self-assembled monolayers of spheres (typically made of Polystyrene) as evaporation masks. This method has been used to fabricate arrays of gold nanodots with precisely controlled spacings.
- **Proton beam writing** uses a focused beam of high energy (MeV) protons to pattern resist material at nanodimensions.
- **Charged-particle lithography**, such as ion- or electron-projection lithographies (PREVAIL, SCALPEL, LEEPL), are also capable of high-resolution patterning. Ion beam lithography uses a focused or broad beam of energetic lightweight ions (like He^+) for transferring pattern to a surface. Using Ion Beam Proximity Lithography (IBL) nano-scale features can be transferred on non-planar surfaces.
- **Neutral particle lithography** (NPL) uses a broad beam of energetic neutral particle for pattern transfer on a surface.

PHOTOLITHOGRAPHY

Photolithography, also termed optical lithography or UV lithography, is a process used in microfabrication to pattern parts of a thin film or the bulk of a substrate. It uses light to transfer a geometric pattern from a photomask to a light-sensitive chemical "photoresist", or simply "resist," on the substrate. A series of chemical treatments then either engraves the exposure pattern into, or enables deposition of a new material in the desired pattern upon, the material underneath the photo resist. For example, in complex integrated circuits, a modern CMOS wafer will go through the photolithographic cycle up to 50 times.

Photolithography shares some fundamental principles with photography in that the pattern in the etching resist is created by exposing it to light, either directly (without using a mask) or with a projected image using an optical mask. This procedure is comparable to a high precision version of the method used to make printed circuit boards. Subsequent stages in the process have more in common with etching than with lithographic printing. It is used because it can create extremely small patterns (down to a few tens of nanometers in size), it affords exact control over the shape and size of the objects it creates, and because it can create patterns over an entire surface cost-effectively. Its main disadvantages are that it requires a flat substrate to start with, it is not very effective at creating shapes that are not flat, and it can require extremely clean operating conditions.

The root words *photo*, *litho*, and *graphy* all have Greek origins, with the meanings 'light', 'stone' and 'writing' respectively. As suggested by the name compounded from them, *photolithography* is a printing method (originally based on the use of limestone printing plates) in which light plays an essential role. In the 1820s, Nicephore Niepce invented a photographic process that used Bitumen of Judea, a natural asphalt, as the first photoresist. A thin coating of the bitumen on a sheet of metal, glass or stone became less soluble where it was exposed to light; the unexposed parts could then be rinsed away with a suitable solvent, baring the material beneath, which was then chemically etched in an acid bath to produce a printing plate. The light-sensitivity of bitumen was very poor and very long exposures were required, but despite the later introduction of more sensitive alternatives, its low cost and superb resistance to strong acids prolonged its commercial life into the early 20th century. In 1940, Oskar Süß created a *positive* photoresist by using diazonaphthoquinone, which worked in the opposite manner: the coating was initially insoluble and was rendered soluble where it was exposed to light. In 1954, Louis Plambeck Jr. developed the Dycryl polymeric letterpress plate, which made the platemaking process faster.

Basic Procedure

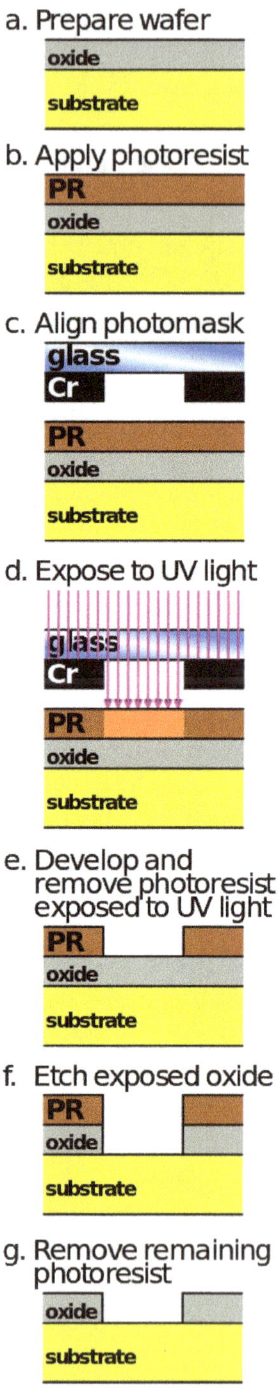

Simplified illustration of dry etching using positive photoresist during a photolithography process in semiconductor microfabrication (not to scale).

A single iteration of photolithography combines several steps in sequence. Modern cleanrooms use automated, robotic wafer track systems to coordinate the process. The procedure described here omits some advanced treatments, such as thinning agents or edge-bead removal.

Cleaning

If organic or inorganic contaminations are present on the wafer surface, they are usually removed by wet chemical treatment, e.g. the RCA clean procedure based on solutions containing hydrogen peroxide. Other solutions made with trichloroethylene, acetone or methanol can also be used to clean.

Preparation

The wafer is initially heated to a temperature sufficient to drive off any moisture that may be present on the wafer surface, 150 °C for ten minutes is sufficient. Wafers that have been in storage must be chemically cleaned to remove contamination. A liquid or gaseous “adhesion promoter”, such as Bis(trimethylsilyl)amine (“hexamethyldisilazane”, HMDS), is applied to promote adhesion of the photoresist to the wafer. The surface layer of silicon dioxide on the wafer reacts with HMDS to form tri-methylated silicon-dioxide, a highly water repellent layer not unlike the layer of wax on a car’s paint. This water repellent layer prevents the aqueous developer from penetrating between the photoresist layer and the wafer’s surface, thus preventing so-called lifting of small photoresist structures in the (developing) pattern. In order to ensure the development of the image, it is best covered and placed over a hot plate and let it dry while stabilizing the temperature at 120 °C.

Photoresist Application

The wafer is covered with photoresist by spin coating. A viscous, liquid solution of photoresist is dispensed onto the wafer, and the wafer is spun rapidly to produce a uniformly thick layer. The spin coating typically runs at 1200 to 4800 rpm for 30 to 60 seconds, and produces a layer between 0.5 and 2.5 micrometres thick. The spin coating process results in a uniform thin layer, usually with uniformity of within 5 to 10 nanometres. This uniformity can be explained by detailed fluid-mechanical modelling, which shows that the resist moves much faster at the top of the layer than at the bottom, where viscous forces bind the resist to the wafer surface. Thus, the top layer of resist is quickly ejected from the wafer’s edge while the bottom layer still creeps slowly radially along the wafer. In this way, any ‹bump’ or ‹ridge’ of resist is removed, leaving a very flat layer. Final thickness is also determined by the evaporation of liquid solvents from the resist. For very small, dense features (< 125 or so nm), lower resist thicknesses (< 0.5 micrometres) are needed to overcome collapse effects at high aspect ratios; typical aspect ratios are < 4:1.

The photo resist-coated wafer is then prebaked to drive off excess photoresist solvent, typically at 90 to 100 °C for 30 to 60 seconds on a hotplate.

Exposure and Developing

After prebaking, the photoresist is exposed to a pattern of intense light. The exposure to light causes a chemical change that allows some of the photoresist to be removed by a special solution, called "developer" by analogy with photographic developer. Positive photoresist, the most common type, becomes soluble in the developer when exposed; with negative photoresist, unexposed regions are soluble in the developer.

A post-exposure bake (PEB) is performed before developing, typically to help reduce standing wave phenomena caused by the destructive and constructive interference patterns of the incident light. In deep ultraviolet lithography, chemically amplified resist (CAR) chemistry is used. This process is much more sensitive to PEB time, temperature, and delay, as most of the "exposure" reaction (creating acid, making the polymer soluble in the basic developer) actually occurs in the PEB.

The develop chemistry is delivered on a spinner, much like photoresist. Developers originally often contained sodium hydroxide (NaOH). However, sodium is considered an extremely undesirable contaminant in MOSFET fabrication because it degrades the insulating properties of gate oxides (specifically, sodium ions can migrate in and out of the gate, changing the threshold voltage of the transistor and making it harder or easier to turn the transistor on over time). Metal-ion-free developers such as tetramethylammonium hydroxide (TMAH) are now used.

The resulting wafer is then "hard-baked" if a non-chemically amplified resist was used, typically at 120 to 180 °Cfor 20 to 30 minutes. The hard bake solidifies the remaining photoresist, to make a more durable protecting layer in future ion implantation, wet chemical etching, or plasma etching.

Etching

In etching, a liquid ("wet") or plasma ("dry") chemical agent removes the uppermost layer of the substrate in the areas that are not protected by photoresist. In semiconductor fabrication, dry etching techniques are generally used, as they can be made anisotropic, in order to avoid significant undercutting of the photoresist pattern. This is essential when the width of the features to be defined is similar to or less than the thickness of the material being etched (i.e. when the aspect ratio approaches unity). Wet etch processes are generally isotropic in nature, which is often indispensable for microelectromechanical systems, where suspended structures must be "released" from the underlying layer.

The development of low-defectivity anisotropic dry-etch process has enabled the ever-smaller features defined photolithographically in the resist to be transferred to the substrate material.

Photoresist Removal

After a photoresist is no longer needed, it must be removed from the substrate. This usually requires a liquid "resist stripper", which chemically alters the resist so that it no longer adheres to the substrate. Alternatively, photoresist may be removed by a plasma containing oxygen, which oxidizes it. This process is called ashing, and resembles dry etching. Use of 1-Methyl-2-pyrrolidone (NMP) solvent for photoresist is another method used to remove an image. When the resist has been dissolved, the solvent can be removed by heating to 80 °C without leaving any residue.by Rex photography and wide angle images

Exposure ("printing") Systems

The wafer track portion of an aligner that uses 365 nm ultraviolet light.

Exposure systems typically produce an image on the wafer using a photomask. The photomask blocks light in some areas and lets it pass in others. (Maskless lithography projects a precise beam directly onto the wafer without using a mask, but it is not widely used in commercial processes.) Exposure systems may be classified by the optics that transfer the image from the mask to the wafer.

Contact and Proximity

A contact printer, the simplest exposure system, puts a photomask in direct contact with the wafer and exposes it to a uniform light. A proximity printer puts a small gap between the photomask and wafer. In both cases, the mask covers the entire wafer, and simultaneously patterns every die.

Contact printing is liable to damage both the mask and the wafer, and this was the primary reason it was abandoned for high volume production. Both contact and proximity lithography require the light intensity to be uniform across an entire wafer, and the mask to align precisely to features already on the wafer. As modern processes use increasingly large wafers, these conditions become increasingly difficult.

Research and prototyping processes often use contact or proximity lithography, because it uses inexpensive hardware and can achieve high optical resolution. The resolution in proximity lithography is approximately the square root of the product of the wavelength and the gap distance. Hence, except for projection lithography (see below), contact printing offers the best resolution, because its gap distance is approximately zero (neglecting the thickness of the photoresist itself). In addition, nanoimprint lithography may revive interest in this familiar technique, especially since the cost of ownership is expected to be low; however, the shortcomings of contact printing discussed above remain as challenges.

Projection

Very-large-scale integration (VLSI) lithography uses projection systems. Unlike contact or proximity masks, which cover an entire wafer, projection masks (known as "reticles") show only one die or an array of dies (known as a "field"). Projection exposure systems (steppers or scanners) project the mask onto the wafer many times to create the complete pattern.

Photomasks

The image for the mask originates from a computerized data file. This data file is converted to a series of polygons and written onto a square fused quartz substrate covered with a layer of chromium using a photolithographic process. A laser beam (laser writer) or a beam of electrons (e-beam writer) is used to expose the pattern defined by the data file and travels over the surface of the substrate in either a vector or raster scan manner. Where the photoresist on the mask is exposed, the chrome can be etched away, leaving a clear path for the illumination light in the stepper/scanner system to travel through.

Resolution in Projection Systems

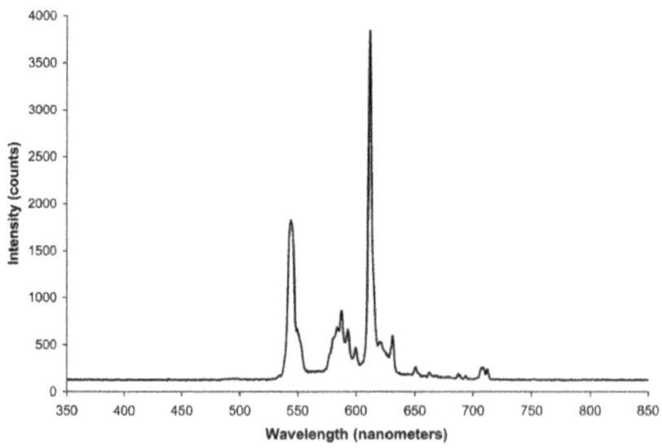

The filtered fluorescent lighting in photolithography cleanrooms contains no ultraviolet or blue light in order to avoid exposing photoresists. The spectrum of light emitted by such fixtures gives virtually all such spaces a bright yellow color.

The ability to project a clear image of a small feature onto the wafer is limited by the wavelength of the light that is used, and the ability of the reduction lens system to capture enough diffraction orders from the illuminated mask. Current state-of-the-art photolithography tools use deep ultraviolet (DUV) light from excimer lasers with wavelengths of 248 and 193 nm (the dominant lithography technology today is thus also called "excimer laser lithography"), which allow minimum feature sizes down to 50 nm. Excimer laser lithography has thus played a critical role in the continued advance of the so-called Moore's Law for the last 20 years (see below).

The minimum feature size that a projection system can print is given approximately by:

$$CD = k_1 \cdot \frac{\lambda}{NA}$$

where

CD is the **minimum feature size** (also called the **critical dimension**, *target design rule*). It is also common to write 2 *times* the *half-pitch*.

k_1 (commonly called *k1 factor*) is a coefficient that encapsulates process-related factors, and typically equals 0.4 for production. The minimum feature size can be reduced by decreasing this coefficient through Computational lithography.

λ is the wavelength of light used

NA is the numerical aperture of the lens as seen from the wafer

According to this equation, minimum feature sizes can be decreased by decreasing the wavelength, and increasing the numerical aperture (to achieve a tighter focused beam and a smaller spot size). However, this design method runs into a competing constraint. In modern systems, the depth of focus is also a concern:

$$D_F = k_2 \cdot \frac{\lambda}{NA^2}$$

Here, k_2 is another process-related coefficient. The depth of focus restricts the thickness of the photoresist and the depth of the topography on the wafer. Chemical mechanical polishing is often used to flatten topography before high-resolution lithographic steps.

Light Sources

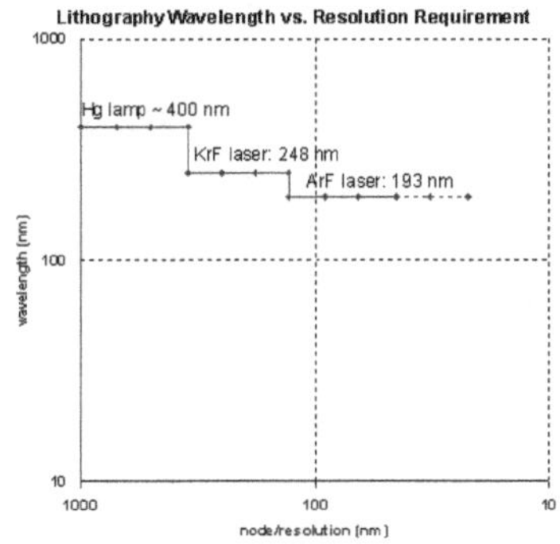

One of the evolutionary paths of lithography has been the use of shorter wavelengths. It is worth noting that the same light source may be used for several technology generations.

Historically, photolithography has used ultraviolet light from gas-discharge lamps using mercury, sometimes in combination with noble gases such as xenon. These lamps produce light across a broad spectrum with several strong peaks in the ultraviolet range. This spectrum is filtered to select a single spectral line. From the early 1960s through the mid-1980s, Hg lamps had been used in lithography for their spectral lines at 436 nm ("g-line"), 405 nm ("h-line") and 365 nm ("i-line"). However, with the semiconductor industry's need for both higher resolution (to produce denser and faster chips) and higher throughput (for lower costs), the lamp-based lithography tools were no longer able to meet the industry's high-end requirements.

This challenge was overcome when in a pioneering development in 1982, excimer laser lithography was proposed and demonstrated at I.B.M. by Kanti Jain, and now excimer laser lithography machines (steppers and scanners) are the primary tools used worldwide in microelectronics production. With phenomenal advances made in tool technology in the last two decades, it is the semiconductor industry view that excimer laser lithography has been a crucial factor in the continued advance of Moore's Law, enabling minimum features sizes in chip manufacturing to shrink from 0.5 micrometer in 1990 to 45 nanometers and below in 2010. This trend is expected to continue into this decade for even denser chips, with minimum features approaching 10 nanometers. From an even broader scientific and technological perspective, in the 50-year history of the laser since its first demonstration in 1960, the invention and development of excimer laser lithography has been recognized as a major milestone.

The commonly used deep ultraviolet excimer lasers in lithography systems are the krypton fluoride laser at 248 nm wavelength and the argon fluoride laser at 193 nm wavelength. The primary manufacturers of excimer laser light sources in the 1980s were Lambda Physik (now part of Coherent, Inc.) and Lumonics. Since the mid-1990s Cymer Inc. has become the dominant supplier of excimer laser sources to the lithography equipment manufacturers, with Gigaphoton Inc. as their closest rival. Generally, an excimer laser is designed to operate with a specific gas mixture; therefore, changing wavelength is not a trivial matter, as the method of generating the new wavelength is completely different, and the absorption characteristics of materials change. For example, air begins to absorb significantly around the 193 nm wavelength; moving to sub-193 nm wavelengths would require installing vacuum pump and purge equipment on the lithography tools (a significant challenge). Furthermore, insulating materials such as silicon dioxide, when exposed to photons with energy greater than the band gap, release free electrons and holes which subsequently cause adverse charging.

Optical lithography has been extended to feature sizes below 50 nm using the 193 nm ArF excimer laser and liquid immersion techniques. Also termed immersion lithography, this enables the use of optics with numerical apertures exceeding 1.0. The liquid used is typically ultra-pure, deionised water, which provides for a refractive index above that of the usual air gap between the lens and the wafer surface. The water is continually circulated to eliminate thermally-induced distortions. Water will only allow *NA*'s of up to ~1.4, but fluids with higher refractive indices would allow the effective *NA* to be increased further.

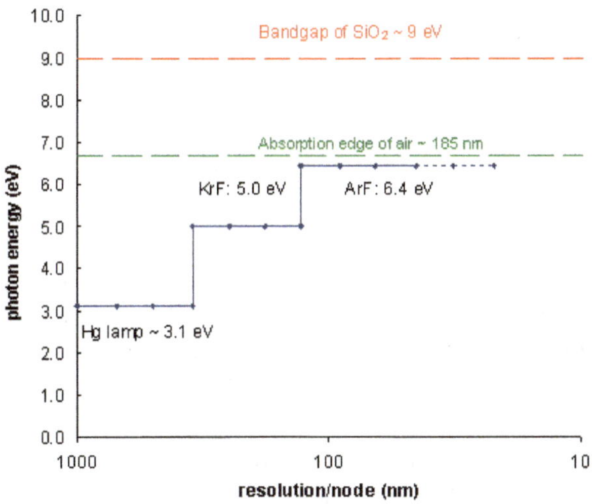

Changing the lithography wavelength is significantly limited by absorption. Air absorbs below ~ 185 nm.

Experimental tools using the 157 nm wavelength from the F2 excimer laser in a manner similar to current exposure systems have been built. These were once targeted to succeed 193 nm lithography at the 65 nm feature size node but have now all but been eliminated by the introduction of immersion lithography. This was due to persistent technical problems with the 157 nm technology and economic considerations that provided strong incentives for the continued use of 193 nm excimer laser lithography technology. High-index immersion lithography is the newest extension of 193 nm lithography to be considered. In 2006, features less than 30 nm were demonstrated by IBM using this technique.

UV excimer lasers have been demonstrated to about 126 nm (for Ar_2*). Mercury arc lamps are designed to maintain a steady DC current of 50 to 150 Volts, however excimer lasers have a higher resolution. Excimer lasers are gas-based light systems that are usually filled with inert and halide gases (Kr, Ar, Xe, F and Cl) that are charged by an electric field. The higher the frequency, the greater the resolution of the image. KrF lasers are able to function at a frequency of 4 kHz . In addition to running at a higher frequency, excimer lasers are compatible with more advanced machines than mercury arc lamps are. They are also able to operate from greater distances (up to 25 meters) and are able to maintain their accuracy with a series of mirrors and antireflective-coated lenses. By setting up multiple lasers and mirrors, the amount of energy loss is minimized, also since the lenses are coated with antireflective material, the light intensity remains relatively the same from when it left the laser to when it hits the wafer.

Lasers have been used to indirectly generate non-coherent extreme UV (EUV) light at 13.5 nm for extreme ultraviolet lithography. The EUV light is not emitted by the laser, but rather by a tin or xenon plasma which is excited by an eximer laser. Fabrication of feature sizes of 10 nm has been demonstrated in production environments, but not yet at rates needed for commercialization. However, this is expected by 2016. This technique does not require a synchrotron, and EUV sources, as noted, do not produce coherent light. However vacuum systems and a number of novel technologies (including much higher EUV energies than are now produced) are needed to work with UV at the edge of the X-ray spectrum (which begins at 10 nm).

Theoretically, an alternative lightsource for photolithography, especially if and when wavelengths continue to decrease to extreme UV or X-ray, is the free-electron laser (or one might say xaser for an X-ray device). Free-electron lasers can produce high quality beams at arbitrary wavelengths.

Experimental Methods

Photolithography has been defeating predictions of its demise for many years. For instance, by the early 1980s, many in the semiconductor industry had come to believe that features smaller than 1 micrometer could not be printed optically. Modern techniques using excimer laser lithography already print features with dimensions a fraction of the wavelength of light used – an amazing optical feat. New tricks such as immersion lithography, dual-tone resist and multiple patterning continue to improve the resolution of 193 nm lithography. Meanwhile, current research is exploring alternatives to conventional UV, such as electron beam lithography, X-ray lithography, extreme ultraviolet lithography and ion projection lithography.

Maskless Lithography

In **maskless lithography**, the radiation that is used to expose a photosensitive emulsion (or photoresist) is not projected from, or transmitted through, a photomask. Instead, most commonly, the radiation is focused to a narrow beam. The beam is then used to directly write the image into the photoresist, one or more pixels at a time. An alternative method, developed by Micronic Laser Systems (today MYCRONIC) or Heidelberg Instruments Mikrotechnik, is to scan a programmable reflective photomask, which is then imaged onto the photoresist. This has the advantage of higher throughput and flexibility. Both methods are used to define patterns on photomasks.

A key advantage of maskless lithography is the ability to change lithography patterns from one run to the next, without incurring the cost of generating a new photomask. This may prove useful for double patterning.

Forms of Maskless Lithography

Currently, the main forms of maskless lithography are electron beam and optical. In addition, focused ion beam systems have established an important niche role in failure analysis and defect repair. Finally, systems based on arrays of probe tips have recently been announced.

Electron Beam

The most commonly used form of maskless lithography today is electron beam lithography. Its widespread use is due to the wide range of electron beam systems available accessing an equally wide range of electron beam energies (~10 eV to ~100 keV). This is already being used in wafer-level production at eASIC, which uses conventional direct-write electron beam lithography to customize a single via layer for low-cost production of ASICs.

Most maskless lithography systems currently being developed are based on the use of multiple electron beams. The goal is to use the parallel scanning of the beams to speed up the patterning of large areas. However, a fundamental consideration here is to what degree electrons from neighboring beams can disturb one another (from Coulomb repulsion). Since the electrons in parallel beams are traveling equally fast, they will persistently repel one another, while the electron lenses act over only a portion of the electrons' trajectories.

Optical

Direct laser writing is a very popular form of optical maskless lithography, which offers flexibility, ease of use, and cost effectiveness in R&D processing. This equipment offers rapid patterning at sub-micrometre resolutions, and offers a compromise between performance and cost when working with feature sizes of approximately 200 nm or greater.

Interference lithography is another form of optical maskless lithography, but is limited to forming periodic patterns only.

For improved image resolution, ultraviolet light, which has a shorter wavelength than visible light, is used to achieve resolution down to around 100 nm. The main optical maskless lithography systems in use today are the ones developed for generating photomasks for the semiconductor and LCD industries.

In 2013, a group at Swinburne University of Technology published their achievement of 9 nm feature size and 52 nm pitch, using a combination of two optical beams of different wavelengths.

Focused ion Beam

Focused ion beam systems are commonly used today for sputtering away defects or uncovering buried features. The use of ion sputtering must take into account the redeposition of sputtered material.

Probe tip Contact

IBM has developed an alternative maskless lithography technique based on atomic force microscopy see here. In addition, Dip Pen Nanolithography is a promising new approach for patterning submicrometre features.

Future of Maskless Lithography

Maskless lithography is already used for the production of photomasks and in limited wafer-level production. There are some obstacles ahead of its use in high-volume manufacturing. First, there is a wide diversity of maskless techniques. Even within the electron-beam category, there are several vendors (Multibeam, Mapper Lithography, Canon, Advantest, Nuflare) with entirely different architectures and beam energies. Second, throughput targets exceeding 10 wafers per hour still need to be met. Third, the capacity and ability to handle the large data volume (Tb-scale) needs to be developed and demonstrated.

In recent years DARPA and NIST have reduced support for maskless lithography in the U.S.

There was a European program that would push the insertion of maskless lithography for IC manufacturing at the 32-nm *half-pitch* node in 2009. Project name was MAGIC, or "MAskless lithoGraphy for IC manufacturing," in frame of EC 7th Framework Programme (FP7).

Due to the increased mask costs for multiple patterning, maskless lithography is once again increasing in visibility.

ION BEAM LITHOGRAPHY

Ion beam lithography is the practice of scanning a focused beam of ions in a patterned fashion across a surface in order to create very small structures such as integrated circuits or other nanostructures.

Ion beam lithography has been found to be useful for transferring high-fidelity patterns on three-dimensional surfaces.

Ion beam lithography offers higher resolution patterning than UV, X-ray, or electron beam lithography because these heavier particles have more momentum. This gives the ion beam a smaller wavelength than even an e-beam and therefore almost no diffraction. The momentum also reduces scattering in the target and in any residual gas. There is also a reduced potential radiation effect to sensitive underlying structures compared to x-ray and e-beam lithography.

Ion beam lithography, or ion projection lithography, is similar to Electron beam lithography, but uses much heavier charged particles, ions. In addition to diffraction being negligible, ions move in straighter paths than electrons do both through vacuum and through matter, so there seems be a potential for very high resolution. Secondary particles (electrons and atoms) have very short range, because of the lower speed of the ions. On the other hand, intense sources are more difficult to make and higher acceleration voltages are needed for a given range. Due to the higher energy loss rate, higher particle energy for a given range and the absence of significant space charge effects, shot noise will tend to be greater.

Fast moving ions interact differently with matter than electrons do, and, due to their higher momentum, their optical properties are different. They have much shorter range in matter and move straighter through it. At low energies, at the end of the range, they lose more of their energy to the atomic nuclei, rather than to the atoms, so that atoms are dislocated rather than ionized. If the ions don't defuse out of the resist, they dope it. The energy loss in matter follows a Bragg curve and has a smaller statistical spread. They are "stiffer" optically, they require larger fields or distances to focus or bend. The higher momentum resists space charge effects.

Collider particle accelerators have shown that it is possible to focus and steer high momentum charged particles with very great precision.

REFERENCES

1. McCord, M. A.; M. J. Rooks (2000). "2". SPIE Handbook of Microlithography, Micromachining and Microfabrication.
2. Parker, N. W.; et al. (2000). "High-throughput NGL electron-beam direct-write lithography system". Proc. SPIE. **3997**: 713. doi:10.1117/12.390042.
3. Faster and lower cost for 65 nm and 45 nm photomask patterning
4. M. L. Kempsell et al., J. Microlith/Nanolith. MEMS MOEMS, vol. 8, 043001(2009).
5. H. Sunaoshi et al., Prof. SPIE vol. 6283, 628306 (2006).
6. K. Ugajin et al., Proc. SPIE vol. 6607, 66070A (2007).
7. F. T. Chen et al., Proc. SPIE vol. 8683, 868311 (2013).
8. L. Feldman; J. Mayer (1986). Fundamentals of Surface and Thin Film Analysis. **54**. pp. 130–133. ISBN 0-444-00989-2.
9. Euronanochem. None. Retrieved on 2011-08-27.
10. Stoffels, E; Stoffels, W W; Kroesen, G M W (2001). "Plasma chemistry and surface processes of negative ions". Plasma Sources Science and Technology. **10** *(2): 311. Bibcode:2001PSST...10..311S. doi:10.1088/0963-0252/10/2/321.*
11. Seah, M. P.; Dench, W. A. (1979). "Quantitative electron spectroscopy of surfaces: A standard data base for electron inelastic mean free paths in solids". Surface and Interface Analysis. **1**: 2. doi:10.1002/sia.740010103.
12. Tanuma, S.; Powell, C. J.; Penn, D. R. (1994). "Calculations of electron inelastic mean free paths. V. Data for 14 organic compounds over the 50–2000 eV range". Surface and Interface Analysis. **21** *(3): 165. doi:10.1002/sia.740210302.*
13. ^ Jump up to:[a] [b] Broers, A. N.; et al. (1996). "Electron beam lithography—Resolution limits". Microelectronic Engineering. **32**: 131–142. doi:10.1016/0167-9317(95)00368-1.
14. K. W. Lee (2009). "Secondary electron generation in electron-beam-irradiated solids:resolution limits to nanolithography". J. Kor. Phys. Soc. **55** *(4): 1720. Bibcode:2009JKPS...55.1720L. doi:10.3938/jkps.55.1720. Archived from the original on 2011-07-22.*
15. SPIE Newsroom: Double exposure makes dense high-resolution diffractive optics. Spie.org (2009-11-03). Retrieved on 2011-08-27.

16. Dapor, M.; et al. (2010). "Monte Carlo modeling in the low-energy domain of the secondary electron emission of polymethylmethacrylate for critical-dimension scanning electron microscopy". J. Micro/Nanolith. MEMS MOEMS. **9**: 023001. doi:10.1117/1.3373517.
17. P. T. Henderson; et al. (1999). "Long-distance charge transport in duplex DNA: The phonon-assisted polaron-like hopping mechanism". Proc. Natl. Acad. Sci. U.S.A. **96** *(15): 8353–8358. Bibcode:1999PNAS...96.8353H. doi:10.1073/pnas.96.15.8353. PMC 17521.* PMID 10411879.
18. H. Seiler (1983). "Secondary electron emission in the scanning electron microscope". J. Appl. Phys. **54** *(11): R1–R18. Bibcode:1983JAP....54R...1S. doi:10.1063/1.332840.*
19. G. Denbeaux et al., 2013 International Workshop on EUV Lithography.
20. J. A. Liddle; et al. (2003). "Resist Requirements and Limitations for Nanoscale Electron-Beam Patterning". Mater. Res. Soc. Symp. Proc. **739** *(19): 19–30.*
21. Ivin, V (2002). "The inclusion of secondary electrons and Bremsstrahlung X-rays in an electron beam resist model". Microelectronic Engineering. 61–62: 343. doi:10.1016/S0167-9317(02)00531-2.
22. Yamazaki, Kenji; Kurihara, Kenji; Yamaguchi, Toru; Namatsu, Hideo; Nagase, Masao (1997). "Novel Proximity Effect Including Pattern-Dependent Resist Development in Electron Beam Nanolithography". Japanese Journal of Applied Physics. **36**: 7552. Bibcode:1997JaJAP..36.7552Y. doi:10.1143/JJAP.36.7552.
23. Renoud, R; Attard, C; Ganachaud, J-P; Bartholome, S; Dubus, A (1998). "Influence on the secondary electron yield of the space charge induced in an insulating target by an electron beam". Journal of Physics: Condensed Matter. **10** *(26): 5821. Bibcode:1998JPCM...10.5821R. doi:10.1088/0953-8984/10/26/010.*
24. J. N. Helbert et al., Macromolecules, vol. 11, 1104 (1978).
25. M. J. Wieland et al., Proc. SPIE vol. 7271, 72710O (2009)
26. F. T. Chen et al., Proc. SPIE vol. 8326, 83262L (2012)
27. P. Kruit et al., J. Vac. Sci. Tech. B 22, 2948 (2004).
28. Bermudez, V. M. (1999). "Low-energy electron-beam effects on poly(methyl methacrylate) resist films". Journal of Vacuum Science and Technology B. **17** *(6): 2512. Bibcode:1999JVSTB..17.2512B. doi:10.1116/1.591134.*
29. H. Yang et al., Proceedings of the 1st IEEE Intl. Conf. on Nano/Micro Engineered and Molecular Systems, pp. 391–394 (2006).
30. Cumming, D. R. S.; Thoms, S.; Beaumont, S. P.; Weaver, J. M. R. (1996). "Fabrication of 3 nm wires using 100 keV electron beam lithography and poly(methyl methacrylate) resist". Applied Physics Letters. **68** *(3,) James Watt Nanofabrication Centre): 322. Bibcode:1996ApPhL..68..322C. doi:10.1063/1.116073.*
31. Manfrinato, Vitor R.; Zhang, Lihua; Su, Dong; Duan, Huigao; Hobbs, Richard G.; Stach, Eric A.; Berggren, Karl K. (2013). "Resolution limits of electron-

beam lithography toward the atomic scale". Nano Lett. **13** *(4): 1555–1558. doi:10.1021/nl304715p.*

32. *C. R. K. Marrian (1992). "Electron-beam lithography with the scanning tunneling microscope". Journal of Vacuum Science and Technology.* ***10*** *(B): 2877–2881. Bibcode:1992JVSTB..10.2877M. doi:10.1116/1.585978.*

33. *T. M. Mayer; et al. (1996). "Field emission characteristics of the scanning tunneling microscope for nanolithography". Journal of Vacuum Science and Technology.* ***14*** *(B): 2438–2444. Bibcode:1996JVSTB..14.2438M. doi:10.1116/1.588751.*

34. *L. S. Hordon; et al. (1993). "Limits of low-energy electron optics". Journal of Vacuum Science and Technology.* ***11*** *(B): 2299–2303. Bibcode:1993JVSTB..11.2299H. doi:10.1116/1.586894.*

35. *Egerton, R. F.; et al. (2004). "Radiation damage in the TEM and SEM". Micron.* ***35*** *(6): 399–409. doi:10.1016/j.micron.2004.02.003. PMID 15120123.*

36. Multibeam Corporation. Multibeamcorp.com (2011-03-04). Retrieved on 2011-08-27.

37. Mapper Lithography. Mapper Lithography (2010-01-18). Retrieved on 2011-08-27.

38. IMS Nanofabrications AG. IMS Nanofabrication AG (2011-12-07). Retrieved on 2012-01-15.

39. M. L. Yu *et al.*, JVST B 23, 2589 (2005).

CHAPTER 6

Nanofiber

Chapter Contents

- **Introduction**
- **Synthesis**
- **Potential Applications**
- **Self-Twisting**
- **Subwavelength-Diameter Optical Fibre**
- **Nanofiber Seeding**
- **Polyaniline Nanofibers**

The nonwoven industry generally considers nanofibers as having a diameter of less than one micron, although the National Science Foundation (NSF) defines nanofibers as having at least one dimension of 100 nanometer (nm) or less. The name derives from the nanometer, a scientific measurement unit representing a billionth of a meter, or three to four atoms wide.

Nanofibers are an exciting new class of material used for several value added applications such as medical, filtration, barrier, wipes, personal care, composite, garments, insulation, and energy storage. Special properties of nanofibers make them suitable for a wide range of applications from medical to consumer products and industrial to high-tech applications for aerospace, capacitors, transistors, drug delivery systems, battery separators, energy storage, fuel cells, and information technology .

INTRODUCTION

Nanofibers are defined as fibers with diameters less than 100 nanometres. In the textile industry, this definition is often extended to include fibers as large as 1000 nm diameter. They can be produced by melt processing, interfacial polymerization, electrospinning, antisolvent-induced polymer precipitation and electrostatic spinning.Carbon nanofibers are graphitized fibers produced by catalytic synthesis.

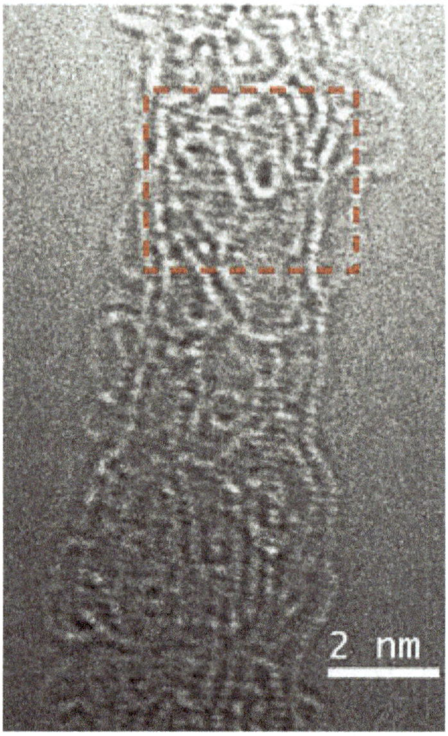

Arrangement of polyvinylidene fluoride molecules in a nanofiber – transmission electron microscopy image.

SYNTHESIS

Inorganic nanofibers can be prepared from various kinds of inorganic substances by several techniques. The most frequently mentioned ceramic materials with nanofiber morphology are titanium dioxide (TiO_2), silicon dioxide (SiO_2), zirconium dioxide (ZrO_2), aluminum oxide (Al_2O_3), lithium titanate ($Li_4Ti_5O_{12}$), titanium nitride (TiN) or platinum (Pt). Production methods include direct drawing from a solution, melt processing, antisolvent-induced polymer precipitation, electrospinning or "island in the sea".

POTENTIAL APPLICATIONS

Nanofibers have applications in medicine, including artificial organ components, tissue engineering, implant material, drug delivery, wound dressing, and medical textile materials. Recently, researchers have found that nanofiber meshes could be used to fight against the HIV-1 virus, and be able to be used as a contraception. In wound healing nanofibers assemble at the injury site and stay put, drawing the body's own growth factors to the injury site. Protective materials include sound absorption materials, protective clothings against chemical and biological warfare agents, and sensor applications for detecting chemical agents. Nanofibers have also been used in pigments for cosmetics.

Applications in the textile industry include sport apparel, sport shoes, climbing, rainwear, outerwear garments, baby diapers.Napkins with nanofibers contain antibodies against numerous biohazards and chemicals that signal by changing color (potentially useful in identifying bacteria in kitchens).

Filtration system applications include HVAC system filters, HEPA, ULPA filters, air, oil, fuel filters for automotive, filters for beverage, pharmacy, medical applications, filter media for new air and liquid filtration applications, such as vacuum cleaners.

Energy applications include Li-ion batteries, photovoltaic cells, membrane fuel cells, and dye-sensitized solar cells. Other applications are micropower to operate personal electronic devices via piezoelectric nanofibers woven into clothing, carrier materials for various catalysts, and photocatalytic air/water purification.

SELF-TWISTING

Self-braiding of nanofibers is related to a balance between flexibility, adhesion, and evaporation of solvent. Its potential applications include: substances that can change optical properties on demand, molecule capture and release for e.g. timed drug delivery, energy storage, and adhesives

SUBWAVELENGTH-DIAMETER OPTICAL FIBRE

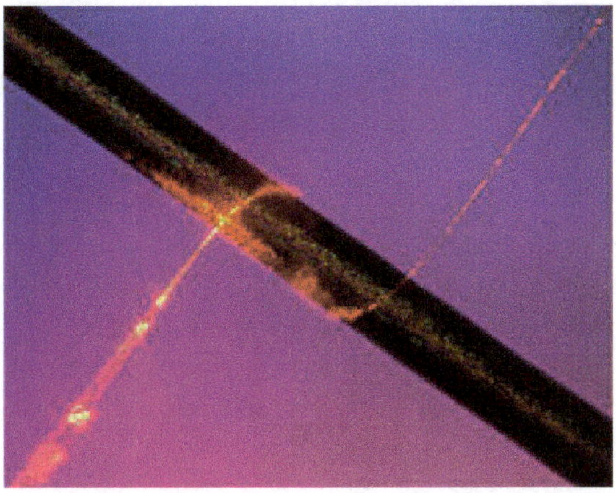

A subwavelength-diameter fibre wraps light around human hair.

A **subwavelength-diameter optical fibre** (**SDF** or **SDOF**) is an optical fibre whose diameter is less than the wavelength of the light being propagated through it. An SDF usually consists of long thick parts (same as conventional optical fibres) at both ends, transition regions (tapers) where the fibre diameter gradually decreases down to the subwavelength value, and a subwavelength-diameter waist, which is the main acting part.

Name

There is no general agreement on how these optical elements are to be named; different groups prefer to emphasize different properties of such fibres, sometimes even using different terms. The names in use include:

- Subwavelength waveguide, subwavelength optical wire, subwavelength-diameter silica wire, subwavelength diameter fibre taper
- (Photonic) wire waveguide, photonic wire, photonic nanowire, optical nanowires, optical fibre nanowires
- Tapered (optical) fibre, fibre taper
- Submicron-diameter silica fibre
- Ultrathin optical fibres
- Optical nanofibre
- Optical microfibres
- Submicron fibre waveguides
- Micro/nano optical wires (MNOW)

The term *waveguide* can be applied not only to fibres, but also to other waveguiding structures such as silicon photonic subwavelength waveguides. The term *submicron* is often synonymous to *subwavelength*, as the majority of experiments are carried out using light with a wavelength between 0.8 and 1.6 μm. All the names with the prefix *nano-* are somewhat misleading, since it is usually applied to objects with dimensions on the scale of nanometers (e.g., nanoparticle, nanotechnology). The characteristic behaviour of the SDF appears when the fibre diameter is about half of the wavelength of light. That is why the term *subwavelength* is the most appropriate for these objects.

Manufacturing

An SDF is usually created by tapering a commercial optical fibre. Special pulling machines accomplish the process. An optical fibre usually consists of a core, a cladding, and a protective coating. Before pulling a fibre, its coating is removed (i.e., the fibre is stripped). The ends of the bare fibre are fixed onto movable "translation" stages on the machine. The middle of the fibre (between the stages) is then heated with a flame or a laser beam; at the same time, the translation stages move in opposite directions. The glass melts and the fibre is elongated, while its diameter decreases.

Using the described method, waists between 1 and 10 nm in length and diameters down to 100 nm are obtained.

Handling

Being extremely thin, an SDF is also extremely fragile. Therefore, an SDF is usually mounted onto a special frame immediately after pulling and is never detached from this frame.

Dust, however, may attach to the surface of an SDF. If significant laser power is coupled into the fibre, the dust particles will scatter light in the evanescent field, heat up, and may thermally destroy the waist. In order to prevent this, SDFs are pulled and used in dust-free environments such as flowboxes or vacuum chambers.

Applications

- Sensors.
- Nonlinear optics.
- Fibre couplers.
- Atom trapping and guiding. See, for example, and.
- Quantum interface for quantum information processing. See, for example, a theoretical analysis with applications to precise quantum nondemolition measurement and all-optical switches.

NANOFIBER SEEDING

Nanofiber Seeding is a process to control the bulk morphology of chemically synthesized conducting polymers. Typically, catalytic amount of nanofiber seeds are added in prior to onset of nanofiber seeding polymerization (reaction), where seeds are served as the 'morphology directing agent' rather than conventional templates (see hard or soft templating methods).

Description

A new synthetic approach, called nanofiber seeding,[1] was developed to control the bulk morphology of chemically synthesized electronic organic polymers. Bulk quantities of nanofibers of conducting polymers such as polyaniline, can be synthesized in one step without the need for conventional templates, surfactants, polymers, or organic solvents.[3]

Conventional oxidative polymerization approaches to nanostructured conducting polymers include the use of hard template zeolites, opals, and controlled pore-size membranes, or soft template such as polymers and surfactants. A "template-free" approach has also been described in which the use of large organic anions results in polyaniline nanofibers and nanotubes having average diameters in the 650-80 nm range.[1]

Standard synthesis of polyaniline yields granular morphology. However, if the conventional reaction is seeded by 1-4 mg (seed quantities) of added pre-synthesized polyaniline nanofibers, (nanofiber seeds could be prepared from interfacial polymerization) the bulk morphology changes dramatically from granular to nano-fibrillar. Furthermore,increased capacitance values were observed in polyaniline nanofibers synthesized by the nanofiber seeding approach. Oxidative polymerization can be also seeded by other nanostructure materials such as vanadium pentoxide nanofibers, where V2O5 nanofibers (i) Rapidly initiate fibrillar polymer growth (ii) Slowly dissolve in aq. 1.0 M HCl, which eliminates template removal steps. Hence only catalytic amounts (4mg) V2O5 nanofibers are needed prior to onset of polymerization, which significantly change the bulk morphology of the polymer precipitate. Moreover, single-walled carbon nanotube and nano fibrous hexapeptide can be also used as templating seeds. This method can be extended to all major classes of conducting polymers, including polypyrrole, PEDOT and other polythiophenes etc. Nanofiber seeding is a convenient approach to obtain thin, substrate-supported, transparent films of nanofibers of conducting polymers without requiring any bulk processing steps.

POLYANILINE NANOFIBERS

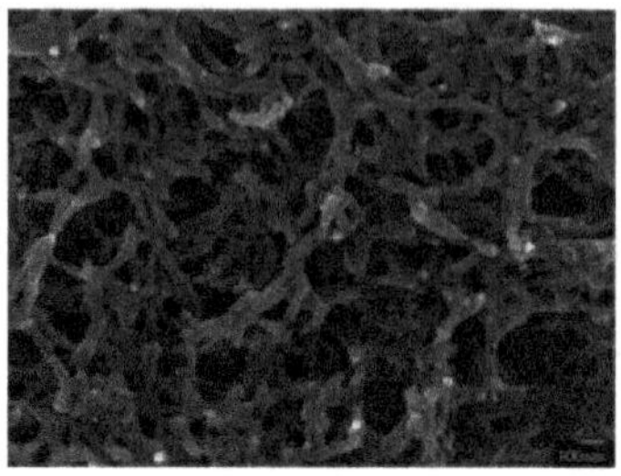

Scanning electron microscope (SEM) image of polyaniline nanofiber film..

Polyaniline nanofibers are a high aspect form of polyaniline, a polymer consisting of aniline monomers, which appears as discrete long threads with an average diameter between 30 nm and 100 nm. Polyaniline is one of the oldest known conducting polymers, being known for over 150 years. Polyaniline nanofibers are often studied for their potential to enhance the properties of polyaniline or have additional beneficial properties due to the addition of a nanostructure to the polymer. Properties that make polyaniline useful can be seen in the nanofiber form as well, such as facile synthesis, environmental stability, and simple acid/base doping/dedoping chemistry. These and other properties have led to the formation of various applications for polyaniline nanofibers as actuators, memory devices, and sensors.

Synthesis

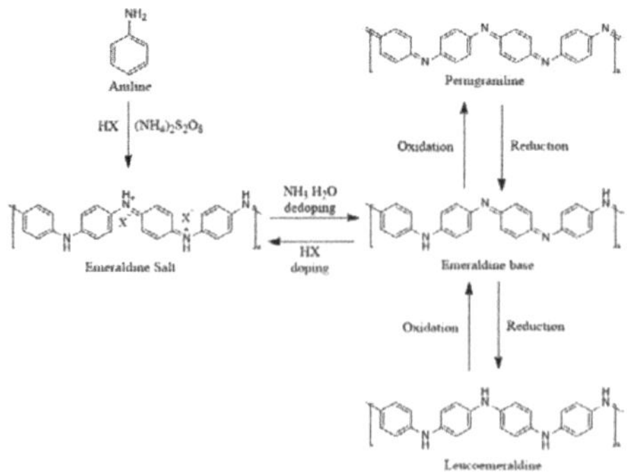

Polymerization pathways of polyaniline and polyaniline nanofibers, as well as the doped/ dedoped oxidation/reduction chemistry that can occur.

Methods for the polymerization of polyaniline nanofibers seen in literature primarily include [redox|chemical oxidative] polymerization, interfacial synthesis, and "rapid mixing" methods. Other less common methods include nanofiber seeding, electrosynthesis, electrospinning, and preforming polymerization in dilute aniline solutions.

Chemical Oxidative Polymerization

Chemical oxidative polymerization is a traditional and commonly used method for the polymerization of aniline in large quantities. When aniline is mixed with an oxidant in an acidic solution, polymerization will occur. The most important parameter to be controlled in this method for the synthesis of polyaniline nanofibers is the domination of homogeneous nucleation over heterogeneous nucleation. Homogeneous nucleation describes when the nuclei are formed spontaneously in solution while heterogeneous nucleation describes when the nuclei are grown on other species. In the early stages of this polymerization, only nanofibers are formed since there are no heteronuclei available for heterogeneous nucleation. However, if the reaction is left uncontrolled, heterogeneous nucleation will begin to dominate as the polyaniline will preferentially grow on existing particles, leading to irreversible agglomeration. The reaction can be made to favor homogeneous nucleation throughout by increasing reaction speed, temperature of the reaction, and allowing the reaction to proceed without stirring.

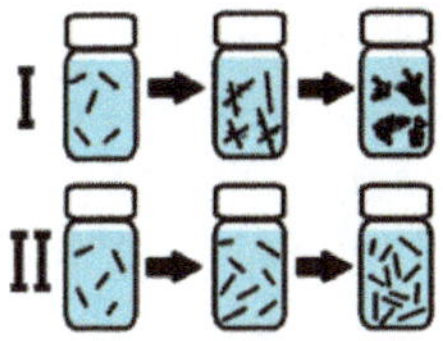

Route I shows the heterogeneous nucleation route, where the nanofibers are formed, followed by secondary growth on the wires which cause agglomeration. Route II shows the homogeneous nucleation route, where only nanofibers are formed.

The diameter of the polyaniline nanofibers can be controlled with this method through choice of acid. Hydrochloric acid produces nanofibers with a diameter of about 30 nm, while camphorsulfonic acid and perchloric acid produce a diameter of 50 nm and 120 nm respectively. Under normal synthetic methods polyaniline derivatives, such as ones that are alkyl and fluoro substituted, do not exhibit a well-defined fibrous shape, however, in the presence of an aniline oligomer nanofibers of certain derivatives can be synthesized. While the most common oxidant is ammonium peroxydisulfate (APS), various others can be used. One study shows the use of potassium biiodate ($KH(IO_3)_2$) as an oxidant, claiming it to lead to polyaniline nanofibers that are longer, have higher crystallinity, and have higher electrical conductivity.

Interfacial Synthesis

In interfacial synthesis, the polymerization happens at the interface between an aqueous and an organic layer. A typical reaction involves an aqueous solution of acid and oxidant and an organic layer of aniline together. This creates the reactive interface for polymerization to occur. As polymerization proceeds, the polyaniline nanofibers will diffuse into the water layer, leaving the reactive interface. This prevents overgrowth onto the existing wires, allowing for homogeneous nucleation to continue occurring. Conditions in the interfacial synthesis can be tuned, such as the type of acid used as well as the oxidant used.

"Rapid Mixing" Reactions

Polyaniline nanofibers can also be synthesized through "rapid mixing" reactions. This method attempts to prevent overgrowth that would compromise the nanofiber nature of the polymer by stopping the polymerization immediately after nanofibers have been formed. This is achieved by the rapid mixing of the monomer, aniline, and an initiator solution. At the start of the reaction, the initiator is consumed rapidly and completely depleted when the nanofibers are formed. Without initiator remaining, the synthesis of polyaniline is halted.

Applications

Monolithic Actuators

Polyaniline nanofibers have been used in the creation of monolithic actuators. They can be used in this application due to their ability to be flash-welded. When exposed to light, polyaniline converts the absorbed energy directly into heat. In a polyaniline film, the heat is dispersed throughout the polymer. In polyaniline nanofibers, however, the heat is trapped within the individual fibers. Therefore, if the intensity of the light is great enough, it will cause the temperature of the nanofibers to rise rapidly, which causes them to weld together or burn. With a moderate flash intensity, the nanofibers will melt rapidly to form a smooth film. Using mask, welds in specific patterns can be made using this technique. In a thick enough sample of nanofibers, only the side exposed to the flash will be welded, creating an asymmetric film where one side remains intact as nanofibers while the other side is effectively crosslinked due to welding. These asymmetric films demonstrate rapid reversible actuation in the presence of acids and bases, in the form of bending and curling. The advantages polyaniline nanofiber asymmetric films have over other actuators include the ease of synthesis, large degree of bending, patternability, and no delamination. These actuators could be used in the development of microtweezers, microvalves, artificial muscles, chemical sensors, and patterned actuator structures.

Digital Memory Devices

Research has shown that polyaniline nanofibers can also be used to create nonvolatile plastic digital memory devices when decorated with various metal, such as gold, nanoparticles. Gold nanoparticles are grown inside dedoped polyaniline nanofibers using a redox reaction. A plastic composite film is placed between two electrodes, and an external bias is used to program ON-OFF states. The switching mechanism is thought to be caused through an interaction between the polyaniline nanofibers and the gold nanoparticles, where charge is transferred to the gold nanoparticles from the polyaniline nanofibers due to an induced electric field. Switching between the ON-OFF states has shown to be rapid, with times of less than 25ns. The retention time of these simple devices are on the order of days after programming, and write-read-erase cycles have been demonstrated.

Chemical Sensing Applications

Polyaniline nanofibers have been shown to be incredibly successful as chemical sensors, as they perform better than conventional polyaniline films in numerous tests. This performance difference has been attributed to their high surface area, porosity, and small diameters which enhance diffusion of materials through the nanofibers. Polyaniline nanofiber sensors function through a change in resistance. The polyaniline nanofiber film is placed on an electrode, where a current flows through. The resistance of the electrode changes when the target interacts with the film, which allows the target to be detected.

One study proposes the creation of hydrogen gas sensors using polyaniline nanofibers. It shows that both doped and dedoped polyaniline nanofibers can be used for detection of hydrogen gas through resistance changes, but the dedoped nanofibers were more stable and had better reproducibility.

Another study shows the potential of polyaniline nanofibers as NO_2 gas sensors. NO_2 gas acts as a strong oxidizing agent to the emeraldine form of polyaniline nanofibers, which causes resistance changes greater than three orders of magnitude at 100 ppm.

Sensing targets can be expanded through adding materials to the polyaniline nanofibers. One study proposes polyaniline nanofiber composites with metal salts for the detection of hydrogen sulfide. Hydrogen sulfide is a weak acid that is dangerous at low ppm, but polyaniline nanofibers can only give a robust response to strong acids. Metal salts can react with hydrogen sulfate to form a metal sulfide precipitate and a strong acid. By combining metal salts and polyaniline nanofibers, detection of hydrogen sulfide can be performed.

Another study decorated polyaniline nanofibers with gold nanoparticles to detect volatile sulfur compounds in expired human breath. These sensors can potentially be used in various breath analyses and also in disease diagnosis for diseases with malodor biomarker gases.

Humidity sensors have also been prepared using polyaniline nanofibers. These sensors were prepared through electrospinning of a N,N-dimethylformamide solution of polyaniline nanofibers, poly(vinyl butyral) (PVB), and poly(ethylene oxide) (PEO). These sensors were shown to have high sensitivity, with resistance changes of three orders of magnitude. Furthermore, the sensors showed good sensing linearity, fast response, small hysterics, and good repeatability.

REFERENCES

1. Lolla, Dinesh; Gorse, Joseph; Kisielowski, Christian; Miao, Jiayuan; Taylor, Philip L.; Chase, George G.; Reneker, Darrell H. (2015). "Polyvinylidene fluoride

molecules in nanofibers, imaged at atomic scale by aberration corrected electron microscopy". Nanoscale. doi:10.1039/C5NR01619C.

2. "Nano glossary". Berlin: Federal Institute for Materials Research and Testing. 24 March 2011. Taken from ISO/TS 27687:2008 Nanotechnologies – Terminology and definitions for nano-objects – Nanoparticle, nanofibre and nanoplate.
3. Zhou, Feng-Lei; Gong, Rong-Hua (2008). "Manufacturing technologies of polymeric nanofibres and nanofibre yarns". Polymer International. 57 (6): 837–845. doi:10.1002/pi.2395. ISSN 0959-8103.
4. Pisignano, Dario (2013). Polymer Nanofibers: Building Blocks for Nanotechnology. Royal Society of Chemistry. pp. 99–. ISBN 978-1-84973-574-2.
5. Xiangwu Zhang, Ph.D., Xiangwu (1 January 2014). Fundamentals of Fiber Science (1st ed.). Lancaster PA: DEStech Publications, Inc. p. 426. ISBN 978-1-60595-119-5. Retrieved 6 February 2015.
6. Nagy ZK, Zsombor K.; Balogh A; Vajna B; Farkas A; Patyi G; Kramarics A; Marosi G (2011). "Comparison of Electrospun and Extruded Soluplus-Based Solid Dosage Forms of Improved Dissolution". Journal of Pharmaceutical Sciences. 101 (1): 322–32. doi:10.1002/jps.22731. PMID 21918982.
7. Pourdeyhimi, Behnam (May 5, 2012) Topic of the month: Commercialization potential of nanofiber textile membranes. Nafigate
8. Nests, Braids And Twists, On The Nano Scale. npr.org. January 9, 2009

Nanotoxicology

Chapter Contents

Nanotoxicology is the study of possibly dangerous effects of elements and even machines at the Nano level. Because elements behave differently and a size of 1-100 Nanometers and this can have a dangerous effect on people and the environment.

Some particles such as carbon Nanotubes that have incredible tensile strength rivaling that of diamonds has long thin Nanofibers that almost exactly resemble the asbestos fibers and if introduced into the lungs can cause a deadly lung condition called mesothelioma (usually caused by asbestos).

INTRODUCTION

Nanotoxicology is the study of the toxicity of nanomaterials. Because of quantum size effects and large surface area to volume ratio, nanomaterials have unique properties compared with their larger counterparts.

Nanotoxicology is a branch of bionanoscience which deals with the study and application of toxicity of nanomaterials. Nanomaterials, even when made of inert elements like gold, become highly active at nanometer dimensions. Nanotoxicological studies are intended to determine whether and to what extent these properties may pose a threat to the environment and to human beings.For instance, Diesel nanoparticles have been found to damage the cardiovascular system in a mouse model.

NANOTOXICOLOGY BACKGROUND

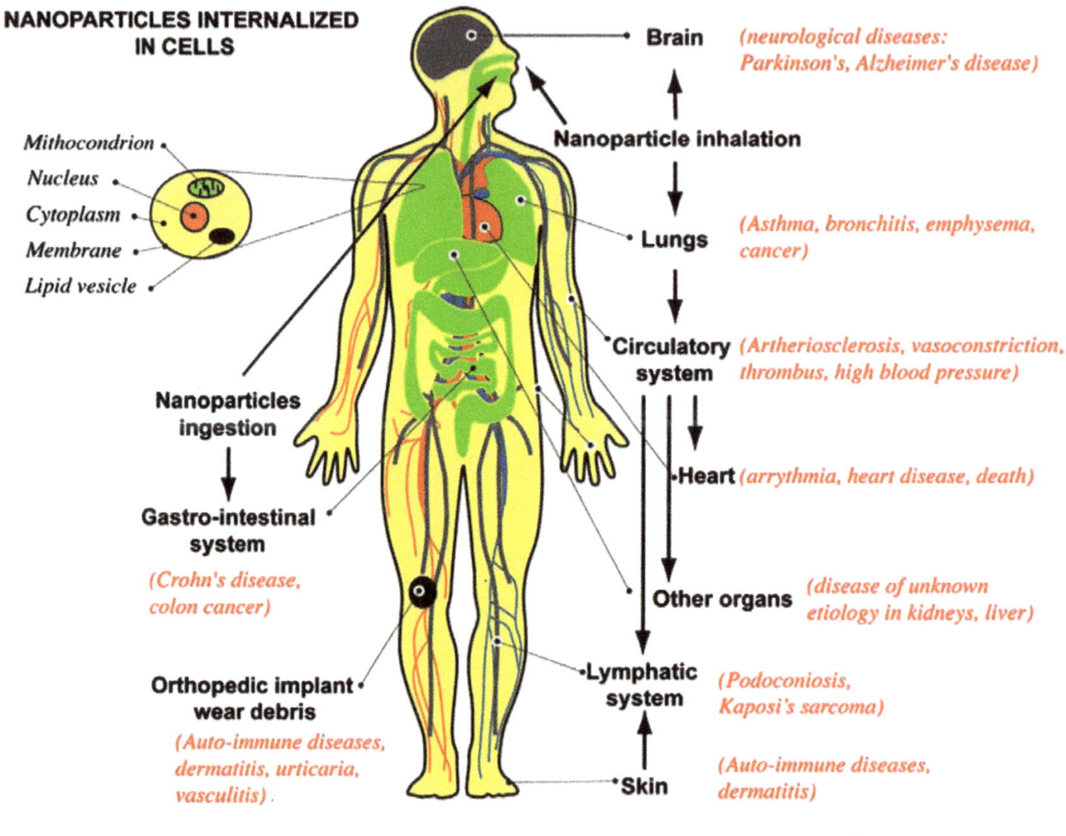

Pathways of exposure to nanoparticles and associated diseases as suggested by epidemiological, in vivo and in vitro studies.

Nanotoxicology is a sub-specialty of particle toxicology. It addresses the toxicology of nanoparticles (particles <100 nm diameter) which appear to have toxicity effects that are unusual and not seen with larger particles. Nanoparticles can be divided into combustion-derived nanoparticles (like diesel soot), manufactured nanoparticles like carbon nanotubes and naturally occurring nanoparticles from volcanic eruptions, atmospheric chemistry etc. Typical nanoparticles that have been studied are titanium dioxide, alumina, zinc oxide, carbon black, and carbon nanotubes, and "nano-C_{60}". Nanoparticles have much larger surface area to unit mass ratios which in some cases may lead to greater pro-inflammatory effects (in, for example, lung tissue). In addition, some nanoparticles seem to be able to translocate from their site of deposition to distant sites such as the blood and the brain. This has resulted in a sea-change in how particle toxicology is viewed- instead of being confined to the lungs, nanoparticle toxicologists study the brain, blood, liver, skin and gut.

Calls for tighter regulation of nanotechnology have arisen alongside a growing debate related to the human health and safety risks associated with nanotechnology. From a large-scale literature review, Yaobo Ding et al. found that release of airborne engineered nanoparticles and associated worker exposure from various production and handling activities at different workplaces are very probable. The Royal Society identifies the potential for nanoparticles to penetrate the skin, and recommends that the use of nanoparticles in cosmetics be conditional upon a favorable assessment by the relevant European Commission safety advisory committee. Andrew Maynard also reports that 'certain nanoparticles may move easily into sensitive lung tissues after inhalation, and cause damage that can lead to chronic breathing problems'.

Carbon nanotubes – characterized by their microscopic size and incredible tensile strength – are frequently likened to asbestos, due to their needle-like fiber shape. In a recent study that introduced carbon nanotubes into the abdominal cavity of mice, results demonstrated that long thin carbon nanotubes showed the same effects as long thin asbestos fibers, raising concerns that exposure to carbon nanotubes may lead to pleural abnormalities such as mesothelioma (cancer of the lining of the lungs caused by exposure to asbestos). Given these risks, effective and rigorous regulation has been called for to determine if, and under what circumstances, carbon nanotubes are manufactured, as well as ensuring their safe handling and disposal.

The Woodrow Wilson Centre's Project on Emerging Technologies conclude that there is insufficient funding for human health and safety research, and as a result there is currently limited understanding of the human health and safety risks associated with nanotechnology. While the US National Nanotechnology Initiative reports that around four percent (about $40 million) is dedicated to risk related research and development, the Woodrow Wilson Centre estimate that only around $11 million is actually directed towards risk related research. They argued in 2007 that it would be necessary to increase funding to a minimum of $50 million in the following two years so as to fill the gaps in knowledge in these areas.

The potential for workplace exposure was highlighted by the 2004 Royal Society report which recommended a review of existing regulations to assess and control workplace exposure to nanoparticles and nanotubes. The report expressed particular concern for the inhalation of large quantities of nanoparticles by workers involved in the manufacturing process.

Stakeholders concerned by the lack of a regulatory framework to assess and control risks associated with the release of nanoparticles and nanotubes have drawn parallels with bovine spongiform encephalopathy ('mad cow's disease'), thalidomide, genetically modified food, nuclear energy, reproductive technologies, biotechnology, and asbestosis. In light of such concerns, the Canadian based ETC Group have called for a moratorium on nano-related research until comprehensive regulatory frameworks are developed that will ensure workplace safety.

Reactive Oxygen Species

For some types of particles, the smaller they are, the greater their surface area to volume ratio and the higher their chemical reactivity and biological activity. The greater chemical reactivity of nanomaterials can result in increased production of reactive oxygen species (ROS), including free radicals. ROS production has been found in a diverse range of nanomaterials including carbon fullerenes, carbon nanotubes and nanoparticle metal oxides. ROS and free radical production is one of the primary mechanisms of nanoparticle toxicity; it may result in oxidative stress, inflammation, and consequent damage to proteins, membranes and DNA.

Biodistribution

The extremely small size of nanomaterials also means that they much more readily gain entry into the human body than larger sized particles. How these nanoparticles behave inside the body is still a major question that needs to be resolved. The behavior of nanoparticles is a function of their size, shape and surface reactivity with the surrounding tissue. In principle, a large number of particles could overload the body's phagocytes, cells that ingest and destroy foreign matter, thereby triggering stress reactions that lead to inflammation and weaken the body's defense against other pathogens. In addition to questions about what happens if non-degradable or slowly degradable nanoparticles accumulate in bodily organs, another concern is their potential interaction or interference with biological processes inside the body. Because of their large surface area, nanoparticles will, on exposure to tissue and fluids, immediately adsorb onto their surface some of the macromolecules they encounter. This may, for instance, affect the regulatory mechanisms of enzymes and other proteins. Nanomaterials are able to cross biological membranes and access cells, tissues and organs that larger-sized particles normally cannot. Nanomaterials can gain access to the blood stream via inhalation or ingestion. At least some nanomaterials can penetrate the skin; even larger microparticles may penetrate skin when it is flexed. Broken skin is an ineffective particle barrier, suggesting that acne, eczema, shaving wounds or severe sunburn may accelerate skin uptake of nanomaterials. Then, once in the blood stream, nanomaterials can be transported around the body and be taken up by organs and tissues, including the brain, heart, liver, kidneys, spleen, bone marrow and nervous system. Nanomaterials have proved toxic to human tissue and cell cultures, resulting in increased oxidative stress, inflammatory cytokine production and cell death. Unlike larger particles, nanomaterials may be taken up by cell mitochondria and the cell nucleus.Studies demonstrate the potential for nanomaterials to cause DNA mutation and induce major structural damage to mitochondria, even resulting in cell death.

Nanotoxicity Studies

There is presently no authority to specifically regulate nanotech-based products. Scientific research has indicated the potential for some nanomaterials to be toxic to humans or the environment. In March 2004 tests conducted by environmental toxicologist Eva Oberdörster, Ph.D. working with Southern Methodist University in Texas, found extensive brain damage to fish exposed to fullerenes for a period of just 48 hours at a relatively moderate dose of 0.5 parts per million (commensurate with levels of other kinds of pollution found in bays). The fish also exhibited changed gene markers in their livers, indicating their entire physiology was affected. In a concurrent test, the fullerenes killed water fleas, an important link in the marine food chain. The extremely small size of fabricated nanomaterials also means that they are much more readily taken up by living tissue than presently known toxins. Nanoparticles can be inhaled, swallowed, absorbed through skin and deliberately or accidentally injected during medical procedures. They might be accidentally or inadvertently released from materials implanted into living tissue. Researcher Shosaku Kashiwada of the National Institute for Environmental Studies in Tsukuba, Japan, in a more recent study, intended to further investigate the effects of nanoparticles on soft-bodied organisms. His study allowed him to explore the distribution of water-suspended fluorescent nanoparticles throughout the eggs and adult bodies of a species of fish, known as the see-through medaka (Oryzias latipes). See-through medaka were used because of their small size, wide temperature and salinity tolerances, and short generation time. Moreover, small fish like the see-through medaka have been popular test subjects for human diseases and organogenesis for other reasons as well, including their transparent embryos, rapid embryo development, and the functional equivalence of their organs and tissue material to that of mammals. Because the see-through medaka have transparent bodies, analyzing the deposition of fluorescent nanoparticles throughout the body is quite simple. For his study, Dr. Kashiwada evaluated four aspects of nanoparticle accumulation. These included the overall accumulation and the size-dependent accumulation of nanoparticles by medaka eggs, the effects of salinity on the aggregation of nanoparticles in solution and on their accumulation by medaka eggs, and the distribution of nanoparticles in the blood and organs of adult medaka. It was also noted that nanoparticles were in fact taken up into the bloodstream and deposited throughout the body. In the medaka eggs, there was a high accumulation of nanoparticles in the yolk; most often bioavailibility was dependent on specific sizes of the particles. Adult samples of medaka had accumulated nanoparticles in the gills, intestine, brain, testis, liver, and bloodstream. One major result from this study was the fact that salinity may have a large influence on the bioavailibility and toxicity of nanoparticles to penetrate membranes and eventually kill the specimen. As the use of nanomaterials increases worldwide, concerns for worker and user safety are mounting. To address such concerns, the Swedish Karolinska Institute conducted a study in which various nanoparticles were introduced to human lung epithelial cells. The results, released in 2008, showed that iron oxide nanoparticles caused little DNA damage and were non-toxic. Zinc oxide nanoparticles were slightly worse. Titanium dioxide caused only DNA damage. Carbon nanotubes caused DNA damage at low levels. Copper oxide was found to be the worst offender, and was the only nanomaterial identified by the researchers as a clear health risk. The latest toxicology studies on mice involving exposure to carbon nanotubes (CNT) showed a limited pulmonary inflammatory potential of MWCNT at levels corresponding to the average inhalable elemental carbon concentrations observed in U.S.-based CNT facilities. The study estimated that considerable years of exposure are necessary for significant pathology to occur.

No Fullerene Toxicity Reported

Nanoparticles can also be made of C_{60}, as is the case with almost any room temperature solid, and several groups have done this and studied toxicity of such particles. The results in the work of Oberdörster at Southern Methodist University, published in "Environmental Health Perspectives" in July 2004, in which questions were raised of potential cytotoxicity, has now been shown by several sources to be likely caused by the tetrahydrofuran used in preparing the 30 nm–100 nm particles of C_{60}used in the research. Isakovic, et al., 2006, who review this phenomenon, gives results showing that removal of THF from the C_{60} particles resulted in a loss of toxicity. Sayes, et al., 2007, also show that particles prepared as in Oberdorster caused no detectable inflammatory response when instilled intratracheally in rats after observation for 3 months, suggesting that even the particles prepared by Oberdorster do not exhibit markers of toxicity in mammalian models. This work used as a benchmark quartz particles, which did give an inflammatory response.

A comprehensive and recent review of work on fullerene toxicity is available in "Toxicity Studies of Fullerenes and Derivatives," a chapter from the book "Bio-applications of Nanoparticles". In this work, the authors review the work on fullerene toxicity beginning in the early 1990s to present, and conclude that the evidence gathered since the discovery of fullerenes overwhelmingly points to C_{60} being non-toxic. As is the case for toxicity profile with any chemical modification of a structural moiety, the authors suggest that individual molecules be assessed individually.

Toxicity of Metal Based Nanoparticles

Metal based nanoparticles (NPs) are a prominent class of NPs synthesized for their functions as semiconductors, electroluminescents, and thermoelectric materials. Biomedically, these antibacterial NPs have been utilized in drug delivery systems to access areas previously inaccessible to conventional medicine. With the recent increase in interest and development of nanotechnology, many studies have been performed to assess whether the unique characteristics of these NPs, namely their small surface area to volume ratio, might negatively impact the environment upon which they were introduced. Researchers have since found that many metal and metal oxide NPs have detrimental effects on the cells with which they come into contact including but not limited to DNA breakage and oxidation, mutations, reduced cell viability, warped morphology, induced apoptosis and necrosis, and decreased proliferation.

Cytotoxicity

A primary marker for the damaging effects of NPs has been cell viability as determined by state and exposed surface area of the cell membrane. Cells exposed to metallic NPs have, in the case of copper oxide, had up to 60% of their cells rendered unviable. When diluted, the positively charged metal ions often experience an electrostatic attraction to the cell membrane of nearby cells, covering the membrane and preventing it from permeating the necessary fuels and wastes. With less exposed membrane for transportation and communication, the cells are often rendered inactive.

NPs have been found to induce apoptosis in certain cells primarily due to the mitochondrial damage and oxidative stress brought on by the foreign NPs electrostatic reactions.

Genotoxicity

Many methods ranging from comet assay to the HPRT gene mutation test have found that metal based NPs disrupt DNA and its replication process in a variety of cells. In a study examining the effects of nanosilver on DNA, AgNPs were introduced to lymphocyte cell DNA which was then examined for abnormalities. The exposure of the NPs correlated to a significant increase in micronuclei indicative of genetic fragmentation. Metal Oxides such as copper oxide, uraninite, and cobalt oxide have also been found to exert significant stress on exposed DNA. The damage done to the DNA will often result in mutated cells and colonies as found with the HPRT gene test.

Coatings and Charges

NPs, in their implementation, are covered with coatings and sometimes given positive or negative charges depending upon the intended function. Studies have found that these external factors affect the degree of toxicity of NPs. Positive charges are usually found to amplify and cause of cellular damage much more noticeably than negative charges do. In a study in which b- and c-polyethylenimine coated AgNPs were attached to strands of Lambda DNA, the cationic b-polyethylenimine coated AgNP was found to lower the melting point of the DNA 50 °C lower than its anionic counterpart.

Immunogenicity of Nanoparticles

Very little attention has been directed towards the potential immunogenicity of nanostructures. Nanostructures can activate the immune system, inducing inflammation, immune responses, allergy, or even affect to the immune cells in a deleterious or beneficial way (immunosuppression in autoimmune diseases, improving immune responses in vaccines). More studies are needed in order to know the potential deleterious or beneficial effects of nanostructures in the immune system. In comparison to conventional pharmaceutical agents, nanostructures have very large sizes, and immune cells, especially phagocytic cells, recognize and try to destroy them.

Complications with nanotoxicity studies

Size is therefore a key factor in determining the potential toxicity of a particle. However it is not the only important factor. Other properties of nanomaterials that influence toxicity include: chemical composition, shape, surface structure, surface charge, aggregation and solubility, and the presence or absence of functional groups of other chemicals. The large number of variables influencing toxicity means that it is difficult to generalise about health risks associated with exposure to nanomaterials – each new nanomaterial must be assessed individually and all material properties must be taken into account. In addition, standarization of toxicology tests between laboratories are needed. Díaz, B. *et al.* from the University of Vigo (Spain) has shown (Small, 2008) that many different cell lines should be studied in order to know if a nanostructure induces toxicity, and human cells can internalize aggregated nanoparticles. Moreover, it is important to take into account that many nanostructures aggregate in biological fluids, but groups manufacturing nanostructures do not care much about this matter. Many efforts of interdisciplinary groups are strongly needed in order to progress in this field.

Effect of Aggregation or Agglomeration of Nanoparticles

Many nanoparticles agglomerate or aggregate when they are placed in environmental or biological fluids. The terms agglomeration and aggregation have distinct definitions according to the standards organizations ISO and ASTM, where agglomeration signifies more loosely bound particles and aggregation signifies very tightly bound or fused particles (typically occurring during synthesis or drying). Nanoparticles frequently agglomerate due to the high ionic strength of environmental and biological fluids, which shields the repulsion due to charges on the nanoparticles. Unfortunately, agglomeration has frequently been ignored in nanotoxicity studies, even though agglomeration would be expected to affect nanotoxicity since it changes the size, surface area, and sedimentation properties of the nanoparticles. In addition, many nanoparticles will agglomerate to some extent in the environment or in the body before they reach their target, so it is desirable to study how toxicity is affected by agglomeration.

A method was published that can be used to produce different mean sizes of stable agglomerates of several metal, metal oxide, and polymer nanoparticles in cell culture media for cell toxicity studies. Different mean sizes of agglomerates are produced by allowing the nanoparticles to agglomerate to a particular size in cell culture media without protein, and then adding protein to coat the agglomerates and "freeze" them at that size. By waiting different amounts of time before adding protein, different mean sizes of agglomerates of a single type of nanoparticle can be produced in an otherwise identical solution, allowing one to study how agglomerate size affects toxicity. In addition, it was found that vortexing while adding a high concentration of nanoparticles to the cell culture media produces much less agglomerated nanoparticles than if the dispersed solution is only mixed after adding the nanoparticles.

The agglomeration/deagglomeration (mechanical stability) potentials of airborne engineered nanoparticle clusters also have significant influences on their size distribution profiles at the end-point of their environmental transport routes. Different aerosolization and deagglomeration systems have been established to test stability of nanoparticle agglomerates. For example, laboratory setups based on critical orifices have been used to apply a wide range of external shear forces onto airborne nanoparticles. After applying shear forces, the particle mean size decreased while the particle generation rate increased. In another pioneering study, four powder aerosolization systems (dustiness testing systems) were compared for the first time for their characteristics linked to aerosol generation.

Challenges of the nano-Visualisation and Related Unknowns in Nanotoxicology

With comparison to more conventional toxicology studies, the nanotoxicology field is however suffering from a lack of easy characterisation of the potential contaminants, the "nano" scale being a scale difficult to comprehend. The biological systems are themselves still not completely known at this scale. Ultimate Atomic visualisation methods such as Electron microscopy (SEM and TEM) and Atomic force microscopy (AFM) analysis allow visualisation of the nano world. Further nanotoxicology studies will require precise characterisation of the specificities of a given nano-element : size, chemical composition, detailed shape, level of aggregation, combination with other vectors, etc. Above all, these properties would have to be determined not only on the nanocomponent before its introduction in the living environment but also in the (mostly aqueous) biological environment.

There is a need for new methodologies to quickly assess the presence and reactivity of nanoparticles in commercial, environmental, and biological samples since current detection techniques require expensive and complex analytical instrumentation. There have been recent attempts to address these issues by developing and investigating sensitive, simple and portable colorimetric detection assays that assess for the surface reactivity of NPs, which can be used to detect the presence of NPs, in environmental and biological relevant samples. Surface redox reactivity is a key emerging property related to potential toxicity of NPs with living cells, and can be used as a key surrogate for determine for the presence of NPs and a first tier analytical strategy toward assessing NP exposures.

It is difficult to determine the degree of effect of a specific nanoparticle when compared to those of comparable nanoparticles already present in our natural environment .

AEM - Analytical Electron Microscopy was used over 40 years ago to investigate amphibole asbestos bodies in Lake Superior from the Reserve Mining operations. This could non-destructively characterise sub-micron particles. Today AEM can fully characterise to atom dimensions.

REFERENCES

1. Cristina Buzea, Ivan Pacheco, and Kevin Robbie "Nanomaterials and Nanoparticles: Sources and Toxicity" Biointerphases 2 (1007) MR17-MR71.
2. Mahmoudi, Morteza; et al. (2012). "Assessing the In Vitro and In Vivo Toxicity of Superparamagnetic Iron Oxide Nanoparticles". Chemical Reviews. 112 (4): 2323–2338. doi:10.1021/cr2002596.
3. http://www.bloomberg.com/apps/news?pid=washingtonstory&sid=aBt.yLf.YfOo study Pollution Particles Lead to Higher Heart Attack Risk (Update1)
4. The Healthy Facilities Institute archive a document exploring the last 63 years of aerosol nanoparticle evaluation at their web site HERE - http://www.healthyfacilitiesinstitute.com/a_253-Commentary_What_You_Cant_See_Can_Still_Hurt_You
5. Ding Y.; et al. (2016). "'Airborne engineered nanomaterials in the workplace—a review of release and worker exposure during nanomaterial production and handling processes". J. Hazard. Mater. doi:10.1016/j.jhazmat.2016.04.075.
6. Andrew Maynard. "Nanotechnology: A Research Strategy for Addressing Risks": 310.
7. Since 2005 a peer reviewed paper, cited by 1058, has remarked that silver nanoparticles are a bactericide and that this property is "only" size dependent, EPA

reference - http://hero.epa.gov/index.cfm?action=reference.details&reference_id=196271

8. Poland C, et al. (2008). "Carbon Nanotubes Introduced into the Abdominal Cavity of Mice Show Asbestos-Like Pathogenicity in a Pilot-Study". Nature Nanotechnology. 3 (7): 423–8. doi:10.1038/nnano.2008.111. PMID 18654567.
9. Woodrow Wilson Centre for International Scholars Project on Emerging Nanotechnologies
10. «An Issues Landscape for Nanotechnology Standards. Report of a Workshop" (PDF). Institute for Food and Agricultural Standards, Michigan State University, East Lansing. 2007.
11. Royal Society and Royal Academy of Engineering (2004). "Nanoscience and nanotechnologies: opportunities and uncertainties". Retrieved 2008-05-18.
12. Rowe G, Horlick-Jones T, Walls J, Pidgeon N (2005). "Difficulties in evaluating public engagement initiatives: reflections on an evaluation of the UK GM Nation?". Public Understanding of Science. 14 (4): 331–352. doi:10.1177/0963662505056611.
13. Nel, Andre; et al. (3 February 2006). "Toxic Potential of Materials at the Nanolevel". Science. 311 (5761): 622–7. doi:10.1126/science.1114397. PMID 16456071.
14. Holsapple, Michael P.; et al. (2005). "Research Strategies for Safety Evaluation of Nanomaterials, Part II: Toxicological and Safety Evaluation of Nanomaterials, Current Challenges and Data Needs". Toxicological Sciences. 88 (1): 12–7. doi:10.1093/toxsci/kfi293. PMID 16120754.
15. Oberdörster, Günter; et al. (2005). "Principles for characterizing the potential human health effects from exposure to nanomaterials: elements of a screening strategy". Particle and Fibre Toxicology. 2: 8. doi:10.1186/1743-8977-2-8. PMC 1260029. PMID 16209704.
16. Hoet, Peter HM; et al. (2004). "Nanoparticles – known and unknown health risks". Journal of Nanobiotechnology. 2 (1): 12. doi:10.1186/1477-3155-2-12. PMC 544578. PMID 15588280.
17. Ryman-Rasmussen, Jessica P.; et al. (2006). "Penetration of Intact Skin by Quantum Dots with Diverse Physicochemical Properties". Toxicological Sciences. 91 (1): 159–65. doi:10.1093/toxsci/kfj122. PMID 16443688.
18. Tinkle, Sally S.; et al. (July 2003). "Skin as a Route of Exposure and Sensitization in Chronic Beryllium Disease". Environmental Health Perspectives. 111 (9): 1202–18. doi:10.1289/ehp.5999.
19. Oberdörster, Günter; et al. (July 2005). "Nanotoxicology: An Emerging Discipline Evolving from Studies of Ultrafine Particles". Environmental Health Perspectives. 113 (7): 823–39. doi:10.1289/ehp.7339. PMC 1257642. PMID 16002369.
20. Li N, Sioutas C, Cho A, et al. (Apr 2003). "Ultrafine particulate pollutants induce oxidative stress and mitochondrial damage". Environ Health Perspect. 111 (4): 455–60. doi:10.1289/ehp.6000. PMC 1241427. PMID 12676598.
21. Porter, Alexandra E.; et al. (2007). "Visualizing the Uptake of C_{60} to the Cytoplasm and Nucleus of Human Monocyte-Derived Macrophage Cells Using Energy-Filtered Transmission Electron Microscopy and Electron Tomography". Environmental Science & Technology. 41 (8): 3012–7. doi:10.1021/es062541f.

22. Geiser, Marianne; et al. (November 2005). "Ultrafine Particles Cross Cellular Membranes by Nonphagocytic Mechanisms in Lungs and in Cultured Cells". Environmental Health Perspectives. 113 (11): 1555–60. doi:10.1289/ehp.8006. PMC 1310918. PMID 16263511.

23. Savic, Radoslav; et al. (25 April 2003). "Micellar Nanocontainers Distribute to Defined Cytoplasmic Organelles". Science. 300 (5619): 615–8. doi:10.1126/science.1078192. PMID 12714738.

24. Kashiwada S (Nov 2006). "Distribution of nanoparticles in the see-through medaka (Oryzias latipes)". Environ Health Perspect. 114 (11): 1697–702. doi:10.1289/ehp.9209. PMC 1665399. PMID 17107855.

25. «Study Sizes up Nanomaterial Toxicity". Chemical & Engineering News. 86 (35). 1 Sep 2008.

26. Aaron Erdely (Oct 2013). "Carbon nanotube dosimetry: from workplace exposure assessment to inhalation toxicology". Particle and Fibre Toxicology. 10.

27. Isakovic A, Markovic Z, Nikolic N, et al. (Oct 2006). "Inactivation of nanocrystalline C_{60} cytotoxicity by gamma-irradiation". Biomaterials. 27 (29): 5049–58. doi:10.1016/j.biomaterials.2006.05.047. PMID 16784774.

28. Sayes CM, Marchione AA, Reed KL, Warheit DB (2007). "Comparative Pulmonary Toxicity Assessments of C_{60} Water Suspensions in Rats: Few Differences in Fullerene Toxicity in Vivo in Contrast to in Vitro Profiles". Nano Lett. 7 (8): 2399–406. doi:10.1021/nl0710710. PMID 17630811.

29. Chan WCW (2007). "Toxicity Studies of Fullerenes and Derivatives". Bio-applications of nanoparticles. New York, NY: Springer Science + Business Media. ISBN 0-387-76712-6.

30. Seabra, Amedea B.; Durán, Nelson (2015-06-03). "Nanotoxicology of Metal Oxide Nanoparticles". Metals. 5 (2): 934–975. doi:10.3390/met5020934.

31. Schrand, Amanda M.; Rahman, Mohammad F.; Hussain, Saber M.; Schlager, John J.; Smith, David A.; Syed, Ali F. (2010-10-01). "Metal-based nanoparticles and their toxicity assessment". Wiley Interdisciplinary Reviews. Nanomedicine and Nanobiotechnology. 2 (5): 544–568. doi:10.1002/wnan.103. ISSN 1939-0041. PMID 20681021.

32. Huk, Anna; Izak-Nau, Emilia; Yamani, Naouale el; Uggerud, Hilde; Vadset, Marit; Zasonska, Beata; Duschl, Albert; Dusinska, Maria (2015-07-24). "Impact of nanosilver on various DNA lesions and HPRT gene mutations – effects of charge and surface coating". Particle and Fibre Toxicology. 12 (1). doi:10.1186/s12989-015-0100-x. PMC 4513976. PMID 26204901.

33. Kim, Jeongeun; Chankeshwara, Sunay V.; Thielbeer, Frank; Jeong, Jiyoung; Donaldson, Ken; Bradley, Mark; Cho, Wan-Seob (2015-05-06). "Surface charge determines the lung inflammogenicity: A study with polystyrene nanoparticles". Nanotoxicology. 10: 1–8. doi:10.3109/17435390.2015.1022887. ISSN 1743-5404. PMID 25946036.

34. Magrez, Arnaud; et al. (2006). "Cellular Toxicity of Carbon-Based Nanomaterials". Nano Letters. 6 (6): 1121–5. doi:10.1021/nl060162e. PMID 16771565.

35. Bharti, Bhuvnesh; et al. (2011). "Aggregation of silica nanoparticles directed by adsorption of Lysozyme". Langmuir. 27: 9823–9833. doi:10.1021/la201898v. PMID 21728288.

36. Zook, Justin; et al. (2011). "Stable nanoparticle aggregates/agglomerates of different

sizes and the effect of their size on hemolytic cytotoxicity". Nanotoxicology. 5: 1–14. doi:10.3109/17435390.2010.536615. PMID 21142841.

37. Ding Y., Riediker M. (2015). "'A System to Assess the Stability of Airborne Nanoparticle Agglomerates Under Aerodynamic Shear". J. Aerosol Sci. 88 (0): 98–108. doi:10.1016/j.jaerosci.2015.06.001.

38. Ding Yaobo, Stahlmecke Burkhard, Sánchez Jiménez Araceli, Tuinman Ilse L., Kaminski Heinz, Kuhlbusch Thomas A. J., Martie , Riediker Michael (2015). "'Dustiness and Deagglomeration Testing: Interlaboratory Comparison of Systems for Nanoparticle Powders". Aerosol Science and Technology. 49 (12): 1222–1231. doi:10.1080/02786826.2015.1114999.

39. http://www.nanosafe.org/scripts/home/publigen/content/templates/show.asp?P=63&L=EN&ITEMID=13

40. Corredor, Charlie; Borysiak, Mark D.; Wolfer, Jay; Westerhoff, Paul; Posner, Jonathan D. (17 March 2015). "Colorimetric Detection of Catalytic Reactivity of Nanoparticles in Complex Matrices". Environmental Science & Technology. 49 (6): 3611–3618. doi:10.1021/es504350j.

41. Mol. Cell. Neurosci. 31 (2006) 642 – 648

CHAPTER 8

Molecular Nanotechnology

Chapter Contents

Molecular nanotechnology (MNT) is a technology based on the ability to build structures to complex, atomic specifications by means of mechanosynthesis.This is distinct from nanoscale materials. Based on Richard Feynman's vision of miniature factories using nanomachines to build complex products (including additional nanomachines), this advanced form of nanotechnology (or *molecular manufacturing*) would make use of positionally-controlled mechanosynthesis guided by molecular machine systems. MNT would involve combining physical principles demonstrated by biophysics, chemistry, other nanotechnologies, and the molecular machinery of life with the systems engineering principles found in modern macroscale factories.

INTRODUCTION

While conventional chemistry uses inexact processes obtaining inexact results, and biology exploits inexact processes to obtain definitive results, molecular nanotechnology would employ original definitive processes to obtain definitive results. The desire in molecular nanotechnology would be to balance molecular reactions in positionally-controlled locations and orientations to obtain desired chemical reactions, and then to build systems by further assembling the products of these reactions.

A roadmap for the development of MNT is an objective of a broadly based technology project led by Battelle (the manager of several U.S. National Laboratories) and the Foresight Institute. The roadmap was originally scheduled for completion by late 2006, but was released in January 2008. The Nanofactory Collaboration is a more focused ongoing effort involving 23 researchers from 10 organizations and 4 countries that is developing a practical research agenda specifically aimed at positionally-controlled diamond mechanosynthesis and diamondoid nanofactory development. In August 2005, a task force consisting of 50+ international experts from various fields was organized by the Center for Responsible Nanotechnology to study the societal implications of molecular nanotechnology.

PROJECTED APPLICATIONS AND CAPABILITIES

Smart Materials and Nanosensors

One proposed Application of MNT is so-called smart materials. This term refers to any sort of material designed and engineered at the nanometer scale for a specific task. It encompasses a wide variety of possible commercial applications. One example would be materials designed to respond differently to various molecules; such a capability could lead, for example, to artificial drugs which would recognize and render inert specific viruses. Another is the idea of self-healing structures, which would repair small tears in a surface naturally in the same way as self-sealing tires or human skin.

A MNT nanosensor would resemble a smart material, involving a small component within a larger machine that would react to its environment and change in some fundamental, intentional way. A very simple example: a photosensor might passively measure the incident light and discharge its absorbed energy as electricity when the light passes above or below a specified threshold, sending a signal to a larger machine. Such a sensor would supposedly cost less and use less power than a conventional sensor, and yet function usefully in all the same applications — for example, turning on parking lot lights when it gets dark.

While smart materials and nanosensors both exemplify useful applications of MNT, they pale in comparison with the complexity of the technology most popularly associated with the term: the replicating nanorobot.

Replicating Nanorobots

MNT nanofacturing is popularly linked with the idea of swarms of coordinated nanoscale robots working together, a popularization of an early proposal by K. Eric Drexler in his 1986 discussions of MNT, but superseded in 1992. In this early proposal, sufficiently capable nanorobots would construct more nanorobots in an artificial environment containing special molecular building blocks.

Critics have doubted both the feasibility of self-replicating nanorobots and the feasibility of control if self-replicating nanorobots could be achieved: they cite the possibility of mutations removing any control and favoring reproduction of mutant pathogenic variations. Advocates address the first doubt by pointing out that the first macroscale autonomous machine replicator, made of Lego blocks, was built and operated experimentally in 2002. While there are sensory advantages present at the macroscale compared to the limited sensorium available at the nanoscale, proposals for positionally controlled nanoscale mechanosynthetic fabrication systems employ dead reckoning of tooltips combined with reliable reaction sequence design to ensure reliable results, hence a limited sensorium is no handicap; similar considerations apply to the positional assembly of small nanoparts. Advocates address the second doubt by arguing that bacteria are (of necessity) evolved to evolve, while nanorobot mutation could be actively prevented by common error-correcting techniques. Similar ideas are advocated in the Foresight Guidelines on Molecular Nanotechnology, and a map of the 137-dimensional replicator design space recently published by Freitas and Merkle provides numerous proposed methods by which replicators could, in principle, be safely controlled by good design.

However, the concept of suppressing mutation raises the question: How can design evolution occur at the nanoscale without a process of random mutation and deterministic selection? Critics argue that MNT advocates have not provided a substitute for such a process of evolution in this nanoscale arena where conventional sensory-based selection processes are lacking. The limits of the sensorium available at the nanoscale could make it difficult or impossible to winnow successes from failures. Advocates argue that design evolution should occur deterministically and strictly under human control, using the conventional engineering paradigm of modeling, design, prototyping, testing, analysis, and redesign.

In any event, since 1992 technical proposals for MNT do not include self-replicating nanorobots, and recent ethical guidelines put forth by MNT advocates prohibit unconstrained self-replication.

Medical Nanorobots

One of the most important applications of MNT would be medical nanorobotics or nanomedicine, an area pioneered by Robert Freitas in numerous books and papers. The ability to design, build, and deploy large numbers of medical nanorobots would, at a minimum, make possible the rapid elimination of disease and the reliable and relatively painless recovery from physical trauma. Medical nanorobots might also make possible the convenient correction of genetic defects, and help to ensure a greatly expanded lifespan. More controversially, medical nanorobots might be used to augment natural human capabilities.

Utility Fog

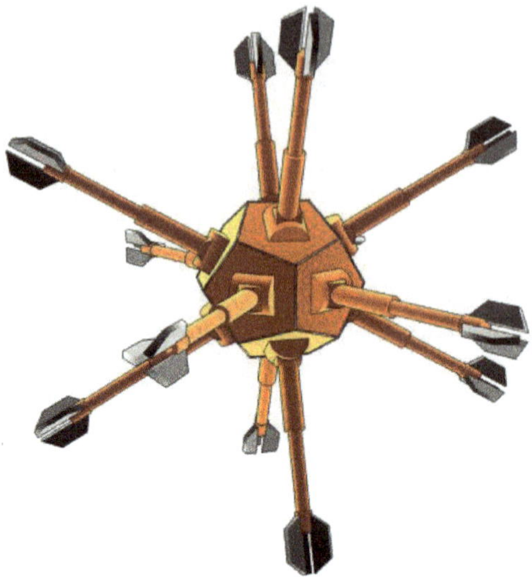

c

Diagram of a 100 micrometer foglet

Another proposed application of molecular nanotechnology is "utility fog" — in which a cloud of networked microscopic robots (simpler than assemblers) would change its shape and properties to form macroscopic objects and tools in accordance with software commands. Rather than modify the current practices of consuming material goods in different forms, utility fog would simply replace many physical objects.

Phased-array Optics

Yet another proposed application of MNT would be phased-array optics (PAO).However, this appears to be a problem addressable by ordinary nanoscale technology. PAO would use the principle of phased-array millimeter technology but at optical wavelengths. This would permit the duplication of any sort of optical effect but virtually. Users could request holograms, sunrises and sunsets, or floating lasers as the mood strikes. PAO systems were described in BC Crandall's *Nanotechnology: Molecular Speculations on Global Abundance* in the Brian Wowk article "Phased-Array Optics."

Potential Social Impacts

Molecular manufacturing is a potential future subfield of nanotechnology that would make it possible to build complex structures at atomic precision. Molecular manufacturing requires significant advances in nanotechnology, but once achieved could produce highly advanced products at low costs and in large quantities in nanofactories weighing a kilogram or more. When nanofactories gain the ability to produce other nanofactories production may only be limited by relatively abundant factors such as input materials, energy and software.

The products of molecular manufacturing could range from cheaper, mass-produced versions of known high-tech products to novel products with added capabilities in many areas of application. Some applications that have been suggested are advanced smart materials, nanosensors, medical nanorobots and space travel. Additionally, molecular manufacturing could be used to cheaply produce highly advanced, durable weapons, which is an area of special concern regarding the impact of nanotechnology. Being equipped with compact computers and motors these could be increasingly autonomous and have a large range of capabilities.

According to Chris Phoenix and Mike Treder from the Center for Responsible Nanotechnology as well as Anders Sandberg from the Future of Humanity Institute molecular manufacturing is the application of nanotechnology that poses the most significant global catastrophic risk. Several nanotechnology researchers state that the bulk of risk from nanotechnology comes from the potential to lead to war, arms races and destructive global government. Several reasons have been suggested why the availability of nanotech weaponry may with significant likelihood lead to unstable arms races (compared to e.g. nuclear arms races): (1) A large number of players may be tempted to enter the race since the threshold for doing so is low; (2) the ability to make weapons with molecular manufacturing will be cheap and easy to hide; (3) therefore lack of insight into the other parties' capabilities can tempt players to arm out of caution or to launch preemptive strikes; (4) molecular manufacturing may reduce dependency on international trade, a potential peace-promoting factor; (5) wars of aggression may pose a smaller economic threat to the aggressor since manufacturing is cheap and humans may not be needed on the battlefield.

Since self-regulation by all state and non-state actors seems hard to achieve, measures to mitigate war-related risks have mainly been proposed in the area of international cooperation. International infrastructure may be expanded giving more sovereignty to the international level. This could help coordinate efforts for arms control. International institutions dedicated specifically to nanotechnology (perhaps analogously to the International Atomic Energy Agency IAEA) or general arms control may also be designed. One may also jointly make differential technological progress on defensive technologies, a policy that players should usually favour. The Center for Responsible Nanotechnology also suggest some technical restrictions.Improved transparency regarding technological capabilities may be another important facilitator for arms-control.

A grey goo is another catastrophic scenario, which was proposed by Eric Drexler in his 1986 book *Engines of Creation*, has been analyzed by Freitas in "Some Limits to Global Ecophagy by Biovorous Nanoreplicators, with Public Policy Recommendations" and has been a theme in mainstream media and fiction. This scenario involves tiny self-replicating robots that consume the entire biosphere using it as a source of energy and building blocks. Nanotech experts including Drexler now discredit the scenario. According to Chris Phoenix a "So-called grey goo could only be the product of a deliberate and difficult engineering process, not an accident". With the advent of nano-biotech, a different scenario called green goo has been forwarded. Here, the malignant substance is not nanobots but rather self-replicating biological organisms engineered through nanotechnology.

Benefits

Nanotechnology (or molecular nanotechnology to refer more specifically to the goals discussed here) will let us continue the historical trends in manufacturing right up to the fundamental limits imposed by physical law. It will let us make remarkably powerful molecular computers. It will let us make materials over fifty times lighter than steel or aluminium alloy but with the same strength. We'll be able to make jets, rockets, cars or even chairs that, by today's standards, would be remarkably light, strong, and inexpensive. Molecular surgical tools, guided by molecular computers and injected into the blood stream could find and destroy cancer cells or invading bacteria, unclog arteries, or provide oxygen when the circulation is impaired.

Nanotechnology will replace our entire manufacturing base with a new, radically more precise, radically less expensive, and radically more flexible way of making products. The aim is not simply to replace today's computer chip making plants, but also to replace the assembly lines for cars, televisions, telephones, books, surgical tools, missiles, bookcases, airplanes, tractors, and all the rest. The objective is a pervasive change in manufacturing, a change that will leave virtually no product untouched. Economic progress and military readiness in the 21st Century will depend fundamentally on maintaining a competitive position in nanotechnology.

Despite the current early developmental status of nanotechnology and molecular nanotechnology, much concern surrounds MNT's anticipated impact on economics and on law. Whatever the exact effects, MNT, if achieved, would tend to reduce the scarcity of manufactured goods and make many more goods (such as food and health aids) manufacturable.

It is generally considered that future citizens of a molecular-nanotechnological society would still need money, in the form of unforgeable digital cash or physical specie (in special circumstances). They might use such money to buy goods and services that are unique, or limited within the solar system. These might include: matter, energy, information, real estate, design services, entertainment services, legal services, fame, political power, or the attention of other people to one's political/religious/philosophical message. Furthermore, futurists must consider war, even between prosperous states, and non-economic goals.

If MNT were realized, some resources would remain limited, because unique physical objects are limited (a plot of land in the real Jerusalem, mining rights to the larger near-earth asteroids) or because they depend on the goodwill of a particular person (the love of a famous person, a live audience in a musical concert). Demand will always exceed supply for some things, and a political economy may continue to exist in any case. Whether the interest in these limited resources would diminish with the advent of virtual reality, where they could be easily substituted, is yet unclear. One reason why it might not is a hypothetical preference for "the real thing", although such an opinion could easily be mollified if virtual reality were to develop to a certain level of quality.

MNT should make possible nanomedical capabilities able to cure any medical condition not already cured by advances in other areas. Good health would be common, and poor health of any form would be as rare as smallpox and scurvy are today. Even cryonics would be feasible, as cryopreserved tissue could be fully repaired.

Risks

Molecular nanotechnology is one of the technologies that some analysts believe could lead to a Technological Singularity. Some feel that molecular nanotechnology would have daunting risks. It conceivably could enable cheaper and more destructive conventional weapons. Also, molecular nanotechnology might permit weapons of mass destruction that could self-replicate, as viruses and cancer cells do when attacking the human body. Commentators generally agree that, in the event molecular nanotechnology were developed, its self-replication should be permitted only under very controlled or "inherently safe" conditions.

A fear exists that nanomechanical robots, if achieved, and if designed to self-replicate using naturally occurring materials (a difficult task), could consume the entire planet in their hunger for raw materials, or simply crowd out natural life, out-competing it for energy (as happened historically when blue-green algae appeared and outcompeted earlier life forms). Some commentators have referred to this situation as the "grey goo" or "ecophagy" scenario. K. Eric Drexler considers an accidental "grey goo" scenario extremely unlikely and says so in later editions of *Engines of Creation*.

In light of this perception of potential danger, the Foresight Institute (founded by K. Eric Drexler to prepare for the arrival of future technologies) has drafted a set of guidelines for the ethical development of nanotechnology. These include the banning of free-foraging self-replicating pseudo-organisms on the Earth's surface, at least, and possibly in other places.

Technical issues and criticism

The feasibility of the basic technologies analyzed in *Nanosystems* has been the subject of a formal scientific review by U.S. National Academy of Sciences, and has also been the focus of extensive debate on the internet and in the popular press.

Study and Recommendations by the U.S. National Academy of Sciences

In 2006, U.S. National Academy of Sciences released the report of a study of molecular manufacturing as part of a longer report, *A Matter of Size: Triennial Review of the National Nanotechnology Initiative* The study committee reviewed the technical content of *Nanosystems*, and in its conclusion states that no current theoretical analysis can be considered definitive regarding several questions of potential system performance, and that optimal paths for implementing high-performance systems cannot be predicted with confidence. It recommends experimental research to advance knowledge in this area:

"Although theoretical calculations can be made today, the eventually attainable range of chemical reaction cycles, error rates, speed of operation, and thermodynamic efficiencies of such bottom-up manufacturing systems cannot be reliably predicted at this time. Thus, the eventually attainable perfection and complexity of manufactured products, while they can be calculated in theory, cannot be predicted with confidence. Finally, the optimum research paths that might lead to systems which greatly exceed the thermodynamic efficiencies and other capabilities of biological systems cannot be reliably predicted at this time. Research funding that is based on the ability of investigators to produce experimental demonstrations that link to abstract models and guide long-term vision is most appropriate to achieve this goal."

Assemblers Versus Nanofactories

A section heading in Drexler's *Engines of Creation* reads "Universal Assemblers", and the following text speaks of multiple types of assemblers which, collectively, could hypothetically "build almost anything that the laws of nature allow to exist." Drexler's colleague Ralph Merkle has noted that, contrary to widespread legend, Drexler never claimed that assembler systems could build absolutely any molecular structure. The endnotes in Drexler's book explain the qualification "almost": "For example, a delicate structure might be designed that, like a stone arch, would self-destruct unless all its pieces were already in place. If there were no room in the design for the placement and removal of a scaffolding, then the structure might be impossible to build. Few structures of practical interest seem likely to exhibit such a problem, however."

In 1992, Drexler published *Nanosystems: Molecular Machinery, Manufacturing, and Computation*, a detailed proposal for synthesizing stiff covalent structures using a table-top factory. Diamondoid structures and other stiff covalent structures, if achieved, would have a wide range of possible applications, going far beyond current MEMS technology. An outline of a path was put forward in 1992 for building a table-top factory in the absence of an assembler. Other researchers have begun advancing tentative, alternative proposed paths for this in the years since Nanosystems was published.

Hard Versus Soft Nanotechnology

In 2004 Richard Jones wrote Soft Machines (nanotechnology and life), a book for lay audiences published by Oxford University. In this book he describes radical nanotechnology (as advocated by Drexler) as a deterministic/mechanistic idea of nano engineered machines that does not take into account the nanoscale challenges such as wetness, stickness, Brownian motion, and high viscosity. He also explains what is soft nanotechnology or more appropriatelly biomimetic nanotechnology which is the way forward, if not the best way, to design functional nanodevices that can cope with all the problems at a nanoscale. One can think of soft nanotechnology as the development of nanomachines that uses the lessons learned from biology on how things work, chemistry to precisely engineer such devices and stochastic physics to model the system and its natural processes in detail.

The Smalley-Drexler Debate

Several researchers, including Nobel Prize winner Dr. Richard Smalley (1943–2005), attacked the notion of universal assemblers, leading to a rebuttal from Drexler and colleagues, and eventually to an exchange of letters. Smalley argued that chemistry is extremely complicated, reactions are hard to control, and that a universal assembler is science fiction. Drexler and colleagues, however, noted that Drexler never proposed universal assemblers able to make absolutely anything, but instead proposed more limited assemblers able to make a very wide variety of things. They challenged the relevance of Smalley's arguments to the more specific proposals advanced in *Nanosystems*. Also, Smalley argued that nearly all of modern chemistry involves reactions that take place in a solvent (usually water), because the small molecules of a solvent contribute many things, such as lowering binding energies for transition states. Since nearly all known chemistry requires a solvent, Smalley felt that Drexler's proposal to use a high vacuum environment was not feasible. However, Drexler addresses this in Nanosystems by showing mathematically that well designed catalysts can provide the effects of a solvent and can fundamentally be made even more efficient than a solvent/enzyme reaction could ever be. It is noteworthy that, contrary to Smalley's opinion that enzymes require water, "Not only do enzymes work vigorously in anhydrous organic media, but in this unnatural milieu they acquire remarkable properties such as greatly enhanced stability, radically altered substrate and enantiomeric specificities, molecular memory, and the ability to catalyse unusual reactions."*Klibanov, AM (April 1989). "Enzymatic catalysis in anhydrous organic solvents.". Trends Biochem Sci.* ***14****: 141–4. doi:10.1016/0968-0004(89)90146-1. PMID 2658221."Zaks, A; Klibanov, AM (April 1989). "Enzymatic catalysis in anhydrous organic solvents" (PDF). Proc. Natl. Acad. Sci. U.S.A.* ***82****: 3192–6. PMC 397741. PMID 3858815.*

Design Issues

For the future, some means have to be found for MNT design evolution at the nanoscale which mimics the process of biological evolution at the molecular scale. Biological evolution proceeds by random variation in ensemble averages of organisms combined with culling of the less-successful variants and reproduction of the more-successful variants, and macroscale engineering design also proceeds by a process of design evolution from simplicity to complexity as set forth somewhat satirically by John Gall: "A complex system that works is invariably found to have evolved from a simple system that worked. . . . A complex system designed from scratch never works and can not be patched up to make it work. You have to start over, beginning with a system that works." A breakthrough in MNT is needed which proceeds from the simple atomic ensembles which can be built with, e.g., an STM to complex MNT systems via a process of design evolution. A handicap in this process is the difficulty of seeing and manipulation at the nanoscale compared to the macroscale which makes deterministic selection of successful trials difficult; in contrast biological evolution proceeds via action of what Richard Dawkins has called the "blind watchmaker" comprising random molecular variation and deterministic reproduction/extinction.

At present in 2007 the practice of nanotechnology embraces both stochastic approaches (in which, for example, supramolecular chemistry creates waterproof pants) and deterministic approaches wherein single molecules (created by stochastic chemistry) are manipulated on substrate surfaces (created by stochastic deposition methods) by deterministic methods comprising nudging them with STM or AFM probes and causing simple binding or cleavage reactions to occur. The dream of a complex, deterministic molecular nanotechnology remains elusive. Since the mid-1990s, thousands of surface scientists and thin film technocrats have latched on to the nanotechnology bandwagon and redefined their disciplines as nanotechnology. This has caused much confusion in the field and has spawned thousands of "nano"-papers on the peer reviewed literature. Most of these reports are extensions of the more ordinary research done in the parent fields.

The Feasibility of the Proposals in *Nanosystems*

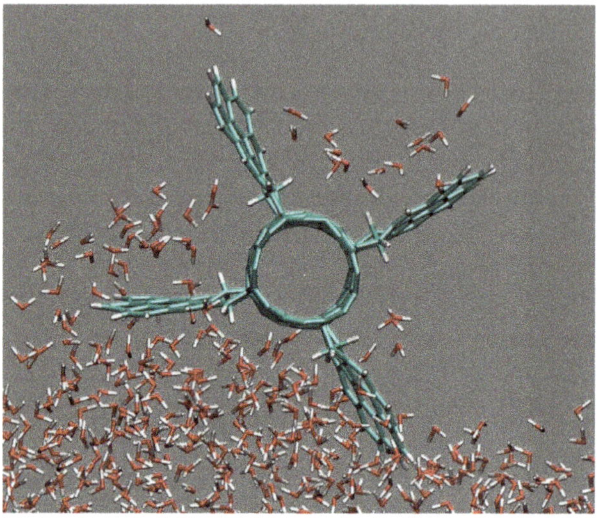

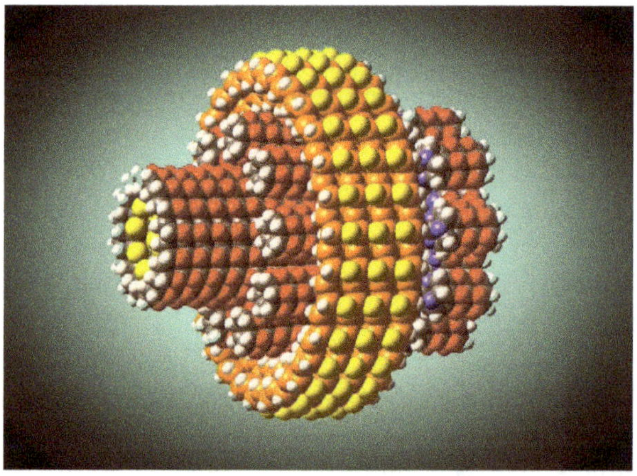

Top, a molecular propellor. Bottom, a molecular planetary gear system. The feasibility of devices like these has been questioned.

The feasibility of Drexler's proposals largely depends, therefore, on whether designs like those in *Nanosystems* could be built in the absence of a universal assembler to build them and would work as described. Supporters of molecular nanotechnology frequently claim that no significant errors have been discovered in *Nanosystems* since 1992. Even some critics concede that "Drexler has carefully considered a number of physical principles underlying the 'high level' aspects of the nanosystems he proposes and, indeed, has thought in some detail" about some issues.

Other critics claim, however, that *Nanosystems* omits important chemical details about the low-level 'machine language' of molecular nanotechnology. They also claim that much of the other low-level chemistry in *Nanosystems* requires extensive further work, and that Drexler's higher-level designs therefore rest on speculative foundations. Recent such further work by Freitas and Merkle is aimed at strengthening these foundations by filling the existing gaps in the low-level chemistry.

Drexler argues that we may need to wait until our conventional nanotechnology improves before solving these issues: "Molecular manufacturing will result from a series of advances in molecular machine systems, much as the first Moon landing resulted from a series of advances in liquid-fuel rocket systems. We are now in a position like that of the British Interplanetary Society of the 1930s which described how multistage liquid-fueled rockets could reach the Moon and pointed to early rockets as illustrations of the basic principle." However, Freitas and Merkle argue that a focused effort to achieve diamond mechanosynthesis (DMS) can begin now, using existing technology, and might achieve success in less than a decade if their "direct-to-DMS approach is pursued rather than a more circuitous development approach that seeks to implement less efficacious nondiamondoid molecular manufacturing technologies before progressing to diamondoid".

To summarize the arguments against feasibility: First, critics argue that a primary barrier to achieving molecular nanotechnology is the lack of an efficient way to create machines on a molecular/atomic scale, especially in the absence of a well-defined path toward a self-replicating assembler or diamondoid nanofactory. Advocates respond that a preliminary research path leading to a diamondoid nanofactory is being developed.

A second difficulty in reaching molecular nanotechnology is design. Hand design of a gear or bearing at the level of atoms might take a few to several weeks. While Drexler, Merkle and others have created designs of simple parts, no comprehensive design effort for anything approaching the complexity of a Model T Ford has been attempted. Advocates respond that it is difficult to undertake a comprehensive design effort in the absence of significant funding for such efforts, and that despite this handicap much useful design-ahead has nevertheless been accomplished with new software tools that have been developed, e.g., at Nanorex.

In the latest report *A Matter of Size: Triennial Review of the National Nanotechnology Initiative* put out by the National Academies Press in December 2006 (roughly twenty years after Engines of Creation was published), no clear way forward toward molecular nanotechnology could yet be seen, as per the conclusion on page 108 of that report: "Although theoretical calculations can be made today, the eventually attainable range of chemical reaction cycles, error rates, speed of operation, and thermodynamic efficiencies of such bottom-up manufacturing systems cannot be reliably predicted at this time. Thus, the eventually attainable perfection and complexity of manufactured products, while they can be calculated in theory, cannot be predicted with confidence. Finally, the optimum research paths that might lead to systems which greatly exceed the thermodynamic efficiencies and other capabilities of biological systems cannot be reliably predicted at this time. Research funding that is based on the ability of investigators to produce experimental demonstrations that link to abstract models and guide long-term vision is most appropriate to achieve this goal." This call for research leading to demonstrations is welcomed by groups such as the Nanofactory Collaboration who are specifically seeking experimental successes in diamond mechanosynthesis. The "Technology Roadmap for Productive Nanosystems" aims to offer additional constructive insights.

It is perhaps interesting to ask whether or not most structures consistent with physical law can in fact be manufactured. Advocates assert that to achieve most of the vision of molecular manufacturing it is not necessary to be able to build "any structure that is compatible with natural law." Rather, it is necessary to be able to build only a sufficient (possibly modest) subset of such structures—as is true, in fact, of any practical manufacturing process used in the world today, and is true even in biology. In any event, as Richard Feynman once said, "It is scientific only to say what's more likely or less likely, and not to be proving all the time what's possible or impossible."

Existing Work on Diamond Mechanosynthesis

There is a growing body of peer-reviewed theoretical work on synthesizing diamond by mechanically removing/adding hydrogen atoms and depositing carbon atoms (a process known as mechanosynthesis). This work is slowly permeating the broader nanoscience community and is being critiqued. For instance, Peng et al. (2006) (in the continuing research effort by Freitas, Merkle and their collaborators) reports that the most-studied mechanosynthesis tooltip motif (DCB6Ge) successfully places a C_2 carbon dimer on a C(110) diamond surface at both 300 K (room temperature) and 80 K (liquid nitrogen temperature), and that the silicon variant (DCB6Si) also works at 80 K but not at 300 K. Over 100,000 CPU hours were invested in this latest study. The DCB6 tooltip motif, initially described by Merkle and Freitas at a Foresight Conference in 2002, was the first complete tooltip ever proposed for diamond mechanosynthesis and remains the only tooltip motif that has been successfully simulated for its intended function on a full 200-atom diamond surface.

The tooltips modeled in this work are intended to be used only in carefully controlled environments (e. g., vacuum). Maximum acceptable limits for tooltip translational and rotational misplacement errors are reported in Peng et al. (2006) -- tooltips must be positioned with great accuracy to avoid bonding the dimer incorrectly. Peng et al. (2006) reports that increasing the handle thickness from 4 support planes of C atoms above the tooltip to 5 planes decreases the resonance frequency of the entire structure from 2.0 THz to 1.8 THz. More importantly, the vibrational footprints of a DCB6Ge tooltip mounted on a 384-atom handle and of the same tooltip mounted on a similarly constrained but much larger 636-atom "crossbar" handle are virtually identical in the non-crossbar directions. Additional computational studies modeling still bigger handle structures are welcome, but the ability to precisely position SPM tips to the requisite atomic accuracy has been repeatedly demonstrated experimentally at low temperature, or even at room temperature constituting a basic existence proof for this capability.

Further research to consider additional tooltips will require time-consuming computational chemistry and difficult laboratory work.

A working nanofactory would require a variety of well-designed tips for different reactions, and detailed analyses of placing atoms on more complicated surfaces. Although this appears a challenging problem given current resources, many tools will be available to help future researchers: Moore's Law predicts further increases in computer power, semiconductor fabrication techniques continue to approach the nanoscale, and researchers grow ever more skilled at using proteins, ribosomes and DNA to perform novel chemistry.

WORKS OF FICTION

- In The Diamond Age by Neal Stephenson diamond can be constructed by simply building it out of carbon atoms. Also all sorts of devices from dust size detection devices to giant diamond zeppelins are constructed atom by atom using only carbon, oxygen, nitrogen and chlorine atoms.
- In the novel *Tomorrow* by Andrew Saltzman (ISBN 1-4243-1027-X), a scientist uses nanorobotics to create a liquid that when inserted into the bloodstream, renders one nearly invincible given that the microscopic machines repair tissue almost instantaneously after it is damaged.
- In the roleplaying game Splicers by Palladium Books, humanity has succumbed to a "nanobot plague" that causes any object made of a non-precious metal to twist and change shape (sometimes into a type of robot) moments after being touched by a human. The object will then proceed to attack the human. This has forced humanity to develop "biotechnological" devices to replace those previously made of metal.
- On the television show Mystery Science Theater 3000, the Nanites (voiced variously by Kevin Murphy, Paul Chaplin, Mary Jo Pehl, and Bridget Jones) - are self-replicating, bio-engineered organisms that work on the ship, they are microscopic creatures that reside in the Satellite of Love's computer systems. (They are similar to the creatures in *Star Trek: The Next Generation* episode "Evolution", which featured "nanites" taking over the *Enterprise*.) The Nanites made their first appearance in season 8. Based on the concept of nanotechnology, their comical *deus ex machina* activities included such diverse tasks as instant repair and construction, hairstyling, performing a Nanite variation of a flea circus, conducting a microscopic war, and even destroying the Observers' planet after a dangerously vague request from Mike to "take care of [a] little problem". They also ran a microbrewery.

MOLECULAR ASSEMBLER

A **molecular assembler**, as defined by K. Eric Drexler, is a "proposed device able to guide chemical reactions by positioning reactive molecules with atomic precision". A molecular assembler is a kind of molecular machine. Some biological molecules such as ribosomes fit this definition. This is because they receive instructions from messenger RNA and then assemble specific sequences of amino acids to construct protein molecules. However, the term "molecular assembler" usually refers to theoretical human-made devices.

Beginning in 2007, the British Engineering and Physical Sciences Research Council has funded development of ribosome-like molecular assemblers. Clearly, molecular assemblers are possible in this limited sense. A technology roadmap project, led by the Battelle Memorial Institute and hosted by several U.S. National Laboratories has explored a range of atomically precise fabrication technologies, including both early-generation and longer-term prospects for programmable molecular assembly; the report was released in December, 2007. In 2008 the Engineering and Physical Sciences Research Council provided funding of 1.5 million pounds over six years for research working towards mechanized mechanosynthesis, in partnership with the Institute for Molecular Manufacturing, amongst others.

Likewise, the term “molecular assembler” has been used in science fiction and popular culture to refer to a wide range of fantastic atom-manipulating nanomachines, many of which may be physically impossible in reality. Much of the controversy regarding “molecular assemblers” results from the confusion in the use of the name for both technical concepts and popular fantasies. In 1992, Drexler introduced the related but better-understood term “molecular manufacturing,” which he defined as the programmed “chemical synthesis of complex structures by mechanically positioning reactive molecules, not by manipulating individual atoms.”

This article mostly discusses “molecular assemblers” in the popular sense. These include hypothetical machines that manipulate individual atoms and machines with organism-like self-replicating abilities, mobility, ability to consume food, and so forth. These are quite different from devices that merely (as defined above) “guide chemical reactions by positioning reactive molecules with atomic precision”.

Because synthetic molecular assemblers have never been constructed and because of the confusion regarding the meaning of the term, there has been much controversy as to whether “molecular assemblers” are possible or simply science fiction. Confusion and controversy also stem from their classification as nanotechnology, which is an active area of laboratory research which has already been applied to the production of real products; however, there had been, until recently, no research efforts into the actual construction of “molecular assemblers”.

Nonetheless, a 2013 paper published in the journal *Science* details a new method of synthesizing a peptide in a sequence-specific manner by using an artificial molecular machine that is guided by a molecular strand. This functions in the same way as a ribosome building proteins by assembling amino acids according to a messenger RNA blueprint. The structure of the machine is based on a rotaxane, which is a molecular ring sliding along a molecular axle. The ring carries a thiolate group which removes amino acids in sequence from the axle, transferring them to a peptide assembly site.

In another paper published in March 2015, also in *Science*, chemists at the University of Illinois report a platform that automates the synthesis of 14 classes of small molecules, with thousands of compatible building blocks.

Nanofactories

A **nanofactory** is a proposed system in which nanomachines (resembling molecular assemblers, or industrial robot arms) would combine reactive molecules via mechanosynthesis to build larger atomically precise parts. These, in turn, would be assembled by positioning mechanisms of assorted sizes to build macroscopic (visible) but still atomically-precise products.

A typical nanofactory would fit in a desktop box, in the vision of K. Eric Drexler published in *Nanosystems: Molecular Machinery, Manufacturing and Computation* (1992), a notable work of "exploratory engineering". During the last decade, others have extended the nanofactory concept, including an analysis of nanofactory convergent assembly by Ralph Merkle, a systems design of a replicating nanofactory architecture by J. Storrs Hall, Forrest Bishop's "Universal Assembler", the patented exponential assembly process by Zyvex, and a top-level systems design for a 'primitive nanofactory' by Chris Phoenix (Director of Research at the Center for Responsible Nanotechnology). All of these nanofactory designs (and more) are summarized in Chapter 4 of *Kinematic Self-Replicating Machines* (2004) by Robert Freitas and Ralph Merkle. The Nanofactory Collaboration, founded by Freitas and Merkle in 2000, is a focused ongoing effort involving 23 researchers from 10 organizations and 4 countries that is developing a practical research agenda specifically aimed at positionally-controlled diamond mechanosynthesis and diamondoid nanofactory development.

In 2005, a computer-animated short film of the nanofactory concept was produced by John Burch, in collaboration with Drexler. Such visions have been the subject of much debate, on several intellectual levels. No one has discovered an insurmountable problem with the underlying theories and no one has proved that the theories can be translated into practice. However, the debate continues, with some of it being summarized in the molecular nanotechnology article.

If nanofactories could be built, severe disruption to the world economy would be one of many possible negative impacts, though it could be argued that this disruption would have little negative effect if everyone had such nanofactories. Great benefits also would be anticipated. Various works of science fiction have explored these and similar concepts. The potential for such devices was part of the mandate of a major UK study led by mechanical engineering professor Dame Ann Dowling.

Self-Replication

"Molecular assemblers" have been confused with self-replicating machines. To produce a practical quantity of a desired product, the nanoscale size of a typical science fiction universal molecular assembler requires an extremely large number of such devices. However, a single such theoretical molecular assembler might be programmed to self-replicate, constructing many copies of itself. This would allow an exponential rate of production. Then after sufficient quantities of the molecular assemblers were available, they would then be re-programmed for production of the desired product. However, if self-replication of molecular assemblers were not restrained then it might lead to competition with naturally occurring organisms. This has been called ecophagy or the grey goo problem.

One method to building molecular assemblers is to mimic evolutionary processes employed by biological systems. Biological evolution proceeds by random variation combined with culling of the less-successful variants and reproduction of the more-successful variants. Production of complex molecular assemblers might be evolved from simpler systems since "A complex system that works is invariably found to have evolved from a simple system that worked. . . . A complex system designed from scratch never works and can not be patched up to make it work. You have to start over, beginning with a system that works."However, most published safety guidelines include "recommendations against developing ... replicator designs which permit surviving mutation or undergoing evolution".

Most assembler designs keep the "source code" external to the physical assembler. At each step of a manufacturing process, that step is read from an ordinary computer file and "broadcast" to all the assemblers. If any assembler gets out of range of that computer, or when the link between that computer and the assemblers is broken, or when that computer is unplugged, the assemblers stop replicating. Such a "broadcast architecture" is one of the safety features recommended by the "Foresight Guidelines on Molecular Nanotechnology", and a map of the 137-dimensional replicator design space recently published by Freitas and Merkle provides numerous practical methods by which replicators can be safely controlled by good design.

Drexler and Smalley Debate

One of the most outspoken critics of some concepts of "molecular assemblers" was Professor Richard Smalley (1943–2005) who won the Nobel prize for his contributions to the field of nanotechnology. Smalley believed that such assemblers were not physically possible and introduced scientific objections to them. His two principal technical objections were termed the "fat fingers problem" and the "sticky fingers problem". He believed these would exclude the possibility of "molecular assemblers" that worked by precision picking and placing of individual atoms. Drexler and coworkers responded to these two issues in a 2001 publication.

Smalley also believed that Drexler's speculations about apocalyptic dangers of self-replicating machines that have been equated with "molecular assemblers" would threaten the public support for development of nanotechnology. To address the debate between Drexler and Smalley regarding molecular assemblers *Chemical & Engineering News* published a point-counterpoint consisting of an exchange of letters that addressed the issues.

Regulation

Speculation on the power of systems that have been called "molecular assemblers" has sparked a wider political discussion on the implication of nanotechnology. This is in part due to the fact that nanotechnology is a very broad term and could include "molecular assemblers." Discussion of the possible implications of fantastic molecular assemblers has prompted calls for regulation of current and future nanotechnology. There are very real concerns with the potential health and ecological impact of nanotechnology that is being integrated in manufactured products. Greenpeace for instance commissioned a report concerning nanotechnology in which they express concern into the toxicity of nanomaterials that have been introduced in the environment. However, it makes only passing references to "assembler" technology. The UK Royal Society and Royal Academy of Engineering also commissioned a report entitled "Nanoscience and nanotechnologies: opportunities and uncertainties" regarding the larger social and ecological implications on nanotechnology. This report does not discuss the threat posed by potential so-called "molecular assemblers."

Formal Scientific Review

In 2006, U.S. National Academy of Sciences released the report of a study of molecular manufacturing as part of a longer report, *A Matter of Size: Triennial Review of the National Nanotechnology Initiative* The study committee reviewed the technical content of *Nanosystems*, and in its conclusion states that no current theoretical analysis can be considered definitive regarding several questions of potential system performance, and that optimal paths for implementing high-performance systems cannot be predicted with confidence. It recommends experimental research to advance knowledge in this area:

"Although theoretical calculations can be made today, the eventually attainable range of chemical reaction cycles, error rates, speed of operation, and thermodynamic efficiencies of such bottom-up manufacturing systems cannot be reliably predicted at this time. Thus, the eventually attainable perfection and complexity of manufactured products, while they can be calculated in theory, cannot be predicted with confidence. Finally, the optimum research paths that might lead to systems which greatly exceed the thermodynamic efficiencies and other capabilities of biological systems cannot be reliably predicted at this time. Research funding that is based on the ability of investigators to produce experimental demonstrations that link to abstract models and guide long-term vision is most appropriate to achieve this goal."

Grey goo

One potential scenario that has been envisioned is out-of-control self-replicating molecular assemblers in the form of grey goo which consumes carbon to continue its replication. If unchecked such mechanical replication could potentially consume whole ecoregions or the whole Earth (ecophagy), or it could simply outcompete natural lifeforms for necessary resources such as carbon, ATP, or UV light (which some nanomotor examples run on). However, the ecophagy and 'grey goo' scenarios, like synthetic molecular assemblers, are based upon still-hypothetical technologies that have not yet been demonstrated experimentally.

In fiction

Molecular assemblers are a popular topic in science fiction, for example, the matter compiler in The Diamond Age and the cornucopia machine in Singularity Sky. The replicator in *Star Trek* might also be considered a molecular assembler. A molecular assembler is also a key element of the plot of the computer game *Deus Ex* (called a "universal constructor" in the game).

In the political sci-fi comic series Transmetropolitan, written by Warren Ellis, machines called "Makers" are used to replicate and reform matter. Each morning, Makers sweep the streets for garbage, gathering the matter to recycle it into more useful objects. The main character also uses a Maker in his apartment to instantly produce a pair of glasses which take photos, as well as other objects such as clothing.

In Dead Money, a DLC of the video game Fallout: New Vegas, the player can obtain useful items from vending machines that use an unknown form of molecular assembly technology to transform casino chips that the player can find into any of several items.

NANOROBOTICS

Nanorobotics is an emerging technology field creating machines or robots which components are at or near the scale of a nanometre (10^{-9} meters).More specifically, nanorobotics (as opposed to microrobotics) refers to the nanotechnology engineering discipline of designing and building nanorobots, with devices ranging in size from 0.1–10 micrometres and constructed of nanoscale or molecular components. The terms *nanobot*, *nanoid*, *nanite*, *nanomachine*, or *nanomite* have also been used to describe such devices currently under research and development.

Nanomachines are largely in the research and development phase, but some primitive molecular machines and nanomotors have been tested. An example is a sensor having a switch approximately 1.5 nanometers across, able to count specific molecules in a chemical sample. The first useful applications of nanomachines may be in nanomedicine. For example, biological machines could be used to identify and destroy cancer cells. Another potential application is the detection of toxic chemicals, and the measurement of their concentrations, in the environment. Rice University has demonstrated a single-molecule car developed by a chemical process and including Buckminsterfullerenes (buckyballs) for wheels. It is actuated by controlling the environmental temperature and by positioning a scanning tunneling microscope tip.

Another definition is a robot that allows precise interactions with nanoscale objects, or can manipulate with nanoscale resolution. Such devices are more related to microscopy or scanning probe microscopy, instead of the description of nanorobots as molecular machine. Using the microscopy definition, even a large apparatus such as an atomic force microscope can be considered a nanorobotic instrument when configured to perform nanomanipulation. For this viewpoint, macroscale robots or microrobots that can move with nanoscale precision can also be considered nanorobots.

Nanorobotics Theory

According to Richard Feynman, it was his former graduate student and collaborator Albert Hibbs who originally suggested to him (circa 1959) the idea of a *medical* use for Feynman's theoretical micromachines (see nanotechnology). Hibbs suggested that certain repair machines might one day be reduced in size to the point that it would, in theory, be possible to (as Feynman put it) "*swallow the surgeon*". The idea was incorporated into Feynman's 1959 essay *There's Plenty of Room at the Bottom.*

Since nanorobots would be microscopic in size, it would probably be necessary for very large numbers of them to work together to perform microscopic and macroscopic tasks. These nanorobot swarms, both those unable to replicate (as in utility fog) and those able to replicate unconstrainedly in the natural environment (as in grey goo and its less common variants, such as synthetic biology or utility fog), are found in many science fiction stories, such as the Borg nanoprobes in *Star Trek* and The Outer Limits episode The New Breed.

Some proponents of nanorobotics, in reaction to the grey goo scenarios that they earlier helped to propagate, hold the view that nanorobots able to replicate outside of a restricted factory environment do not form a necessary part of a purported productive nanotechnology, and that the process of self-replication, were it ever to be developed, could be made inherently safe. They further assert that their current plans for developing and using molecular manufacturing do not in fact include free-foraging replicators.

The most detailed theoretical discussion of nanorobotics, including specific design issues such as sensing, power communication, navigation, manipulation, locomotion, and onboard computation, has been presented in the medical context of nanomedicine by Robert Freitas. Some of these discussions remain at the level of unbuildable generality and do not approach the level of detailed engineering.

Approaches

Biochip

Main article: Biochip

The joint use of nanoelectronics, photolithography, and new biomaterials provides a possible approach to manufacturing nanorobots for common medical uses, such as surgical instrumentation, diagnosis, and drug delivery. This method for manufacturing on nanotechnology scale is in use in the electronics industry since 2008. So, practical nanorobots should be integrated as nanoelectronics devices, which will allow tele-operation and advanced capabilities for medical instrumentation.

Nubots

A *nucleic acid robot* (nubot) is an organic molecular machine at the nanoscale. DNA structure can provide means to assemble 2D and 3D nanomechanical devices. DNA based machines can be activated using small molecules, proteins and other molecules of DNA. Biological circuit gates based on DNA materials have been engineered as molecular machines to allow in-vitro drug delivery for targeted health problems. Such material based systems would work most closely to smart biomaterial drug system delivery, while not allowing precise in vivo teleoperation of such engineered prototypes.

Surface-bound Systems

Several reports have demonstrated the attachment of synthetic molecular motors to surfaces. These primitive nanomachines have been shown to undergo machine-like motions when confined to the surface of a macroscopic material. The surface anchored motors could potentially be used to move and position nanoscale materials on a surface in the manner of a conveyor belt.

Positional Nanoassembly

Nanofactory Collaboration, founded by Robert Freitas and Ralph Merkle in 2000 and involving 23 researchers from 10 organizations and 4 countries, focuses on developing a practical research agenda specifically aimed at developing positionally-controlled diamond mechanosynthesis and a diamondoid nanofactory that would have the capability of building diamondoid medical nanorobots.

Bacteria-Based

This approach proposes the use of biological microorganisms, like the bacterium *Escherichia coli* and *Salmonella typhimurium*. Thus the model uses a flagellum for propulsion purposes. Electromagnetic fields normally control the motion of this kind of biological integrated device. Chemists at the University of Nebraska have created a humidity gauge by fusing a bacterium to a silicone computer chip.

Virus-Based

Retroviruses can be retrained to attach to cells and replace DNA. They go through a process called reverse transcription to deliver genetic packaging in a vector. Usually, these devices are Pol – Gag genes of the virus for the Capsid and Delivery system. This process is called retroviral gene therapy, having the ability to re-engineer cellular DNA by usage of viralvectors. This approach has appeared in the form of retroviral, adenoviral, and lentiviral gene delivery systems. These gene therapy vectors have been used in cats to send genes into the genetically modified organism (GMO), causing it to display the trait.

Open Technology

A document with a proposal on nanobiotech development using open design technology methods, as in open-source hardware and open-source software, has been addressed to the United Nations General Assembly. According to the document sent to the United Nations, in the same way that open source has in recent years accelerated the development of computer systems, a similar approach should benefit the society at large and accelerate nanorobotics development. The use of nanobiotechnology should be established as a human heritage for the coming generations, and developed as an open technology based on ethical practices for peaceful purposes. Open technology is stated as a fundamental key for such an aim.

Nanorobot Race

In the same ways that technology research and development drove the space race and nuclear arms race, a race for nanorobots is occurring. There is plenty of ground allowing nanorobots to be included among the emerging technologies. Some of the reasons are that large corporations, such as General Electric, Hewlett-Packard, Synopsys, Northrop Grumman and Siemens have been recently working in the development and research of nanorobots;surgeons are getting involved and starting to propose ways to apply nanorobots for common medical procedures;universities and research institutes were granted funds by government agencies exceeding $2 billion towards research developing nanodevices for medicine; bankers are also strategically investing with the intent to acquire beforehand rights and royalties on future nanorobots commercialisation. Some aspects of nanorobot litigation and related issues linked to monopoly have already arisen. A large number of patents has been granted recently on nanorobots, done mostly for patent agents, companies specialized solely on building patent portfolios, and lawyers. After a long series of patents and eventually litigations, see for example the Invention of Radio, or the War of Currents, emerging fields of technology tend to become a monopoly, which normally is dominated by large corporations.

Manufacture

3D Printing

3D printing is the process by which a three-dimensional structure is built through the various processes of additive manufacturing. Nanoscale 3D printing involves many of the same process, incorporated at a much smaller scale. To print a structure in the 5-400 μm scale, the precision of the 3D printing machine is improved greatly.

3D printing and Laser Etching

A method pioneered in Seoul, South Korea uses a two-step process of 3D printing, using a 3D printing and laser etched plates. To be more precise at a nanoscale, the 3D printing process uses a laser etching machine, which etches into each plate the details needed for the segment of nanorobot. The plate is then transferred to the 3D printer, which fills the etched regions with the desired nanoparticle. The 3D printing process is repeated until the nanorobot is built from the bottom up. This 3D printing process has many benefits. First, it increases the overall accuracy of the printing process. Second, it has the potential to create functional segments of a nanorobot.

Two-photon Lithography

The 3D printer uses a liquid resin, which is hardened at precisely the correct spots by a focused laser beam. The focal point of the laser beam is guided through the resin by movable mirrors and leaves behind a hardened line of solid polymer, just a few hundred nanometers wide. This fine resolution enables the creation of intricately structured sculptures as tiny as a grain of sand. This process takes place by using photoactive resins, which are hardened by the laser at an extremely small scale to create the structure. This process is quick by nanoscale 3D printing standards. Ultra-small features can be made with the 3D micro-fabrication technique used in multiphoton photopolymerisation. This approach uses a focused laser to trace the desired 3D object into a block of gel. Due to the nonlinear nature of photo excitation, the gel is cured to a solid only in the places where the laser was focused while the remaining gel is then washed away. Feature sizes of under 100 nm are easily produced, as well as complex structures with moving and interlocked parts.

Potential Uses

Nanomedicine

Potential uses for nanorobotics in medicine include early diagnosis and targeted drug-delivery for cancer, biomedical instrumentation, surgery, pharmacokinetics, monitoring of diabetes, and health care.

In such plans, future medical nanotechnology is expected to employ nanorobots injected into the patient to perform work at a cellular level. Such nanorobots intended for use in medicine should be non-replicating, as replication would needlessly increase device complexity, reduce reliability, and interfere with the medical mission.

Nanotechnology provides a wide range of new technologies for developing customized means to optimize the delivery of pharmaceutical drugs. Today, harmful side effects of treatments such as chemotherapy are commonly a result of drug delivery methods that don't pinpoint their intended target cells accurately. Researchers at Harvard and MIT, however, have been able to attach special RNA strands, measuring nearly 10 nm in diameter, to nanoparticles, filling them with a chemotherapy drug. These RNA strands are attracted to cancer cells. When the nanoparticle encounters a cancer cell, it adheres to it, and releases the drug into the cancer cell. This directed method of drug delivery has great potential for treating cancer patients while avoiding negative effects (commonly associated with improper drug delivery). The first demonstration of nanomotors operating in living organism was carried out in 2014 at University of California, San Diego. MRI-guided nanocapsules are one potential precursor to nanorobots.

Another useful application of nanorobots is assisting in the repair of tissue cells alongside white blood cells. Recruiting inflammatory cells or white blood cells (which include neutrophil granulocytes, lymphocytes, monocytes, and mast cells) to the affected area is the first response of tissues to injury. Because of their small size, nanorobots could attach themselves to the surface of recruited white cells, to squeeze their way out through the walls of blood vessels and arrive at the injury site, where they can assist in the tissue repair process. Certain substances could possibly be used to accelerate the recovery.

The science behind this mechanism is quite complex. Passage of cells across the blood endothelium, a process known as transmigration, is a mechanism involving engagement of cell surface receptors to adhesion molecules, active force exertion and dilation of the vessel walls and physical deformation of the migrating cells. By attaching themselves to migrating inflammatory cells, the robots can in effect "hitch a ride" across the blood vessels, bypassing the need for a complex transmigration mechanism of their own.

As of 2016, in the United States, Food and Drug Administration (FDA) regulates nanotechnology on the basis of size.

MECHANOSYNTHESIS

Mechanosynthesis is a term for hypothetical chemical syntheses in which reaction outcomes are determined by the use of mechanical constraints to direct reactive molecules to specific molecular sites. There are presently no chemical syntheses which achieve this aim. Some atomic placement has been achieved with scanning tunnelling microscopes.

In conventional chemical synthesis or chemosynthesis, reactive molecules encounter one another through random thermal motion in a liquid or vapor. In a hypothesized process of mechanosynthesis, reactive molecules would be attached to molecular mechanical systems, and their encounters would result from mechanical motions bringing them together in planned sequences, positions, and orientations. It is envisioned that mechanosynthesis would avoid unwanted reactions by keeping potential reactants apart, and would strongly favor desired reactions by holding reactants together in optimal orientations for many molecular vibration cycles. In biology, the ribosome provides an example of a programmable mechanosynthetic device.

A primitive, very non-biological form of mechanochemistry has been performed at cryogenic temperatures using scanning tunneling microscopes. So far, such devices provide the closest approach to fabrication tools for molecular engineering. Broader exploitation of mechanosynthesis awaits more advanced technology for constructing molecular machine systems, with ribosome-like systems as an attractive early objective.

Much of the excitement regarding advanced mechanosynthesis regards its potential use in assembly of molecular-scale devices. Such techniques appear to have many applications in medicine, aviation, resource extraction, manufacturing and warfare.

Most theoretical explorations of advanced machines of this kind have focused on using carbon, because of the many strong bonds it can form, the many types of chemistry these bonds permit, and utility of these bonds in medical and mechanical applications. Carbon forms diamond, for example, which if cheaply available, would be an excellent material for many machines.

It has been suggested, notably by K. Eric Drexler, that mechanosynthesis will be fundamental to molecular manufacturing based on nanofactories capable of building macroscopic objects with atomic precision. The potential for these has been disputed, notably by Nobel Laureate Richard Smalley (who proposed and then critiqued an unworkable approach based on small fingers).

The Nanofactory Collaboration, founded by Robert Freitas and Ralph Merkle in 2000, is a focused ongoing effort involving 23 researchers from 10 organizations and 4 countries that is developing a practical research agenda specifically aimed at positionally controlled diamond mechanosynthesis and diamondoid nanofactory development.

In practice, getting exactly one molecule to a known place on the microscope's tip is possible, but has proven difficult to automate. Since practical products require at least several hundred million atoms, this technique has not yet proven practical in forming a real product.

The goal of one line of mechanoassembly research focuses on overcoming these problems by calibration, and selection of appropriate synthesis reactions. Some suggest attempting to develop a specialized, very small (roughly 1,000 nanometers on a side) machine tool that can build copies of itself using mechanochemical means, under the control of an external computer. In the literature, such a tool is called an assembler or molecular assembler. Once assemblers exist, geometric growth (directing copies to make copies) could reduce the cost of assemblers rapidly. Control by an external computer should then permit large groups of assemblers to construct large, useful projects to atomic precision. One such project would combine molecular-level conveyor belts with permanently mounted assemblers to produce a factory.

In part to resolve this and related questions about the dangers of industrial accidents and popular fears of runaway events equivalent to Chernobyl and Bhopal disasters, and the more remote issue of ecophagy, grey goo and green goo (various potential disasters arising from runaway replicators, which could be built using mechanosynthesis) the UK Royal Society and UK Royal Academy of Engineering in 2003 commissioned a study to deal with these issues and larger social and ecological implications, led by mechanical engineering professor Ann Dowling. This was anticipated by some to take a strong position on these problems and potentials — and suggest any development path to a general theory of so-called mechanosynthesis. However, the Royal Society's nanotech report did not address molecular manufacturing at all, except to dismiss it along with grey goo.

Current technical proposals for nanofactories do not include self-replicating nanorobots, and recent ethical guidelines would prohibit development of unconstrained self-replication capabilities in nanomachines.

Diamond Mechanosynthesis

There is a growing body of peer-reviewed theoretical work on synthesizing diamond by mechanically removing/adding hydrogen atoms and depositing carbon atoms (a process known as diamond mechanosynthesis or DMS). For example, the 2006 paper in this continuing research effort by Freitas, Merkle and their collaborators reports that the most-studied mechanosynthesis tooltip motif (DCB6Ge) successfully places a C_2 carbon dimer on a C(110) diamond surface at both 300 K (room temperature) and 80 K (liquid nitrogen temperature), and that the silicon variant (DCB6Si) also works at 80 K but not at 300 K. These tooltips are intended to be used only in carefully controlled environments (e.g., vacuum). Maximum acceptable limits for tooltip translational and rotational misplacement errors are reported in paper III—tooltips must be positioned with great accuracy to avoid bonding the dimer incorrectly. Over 100,000 CPU hours were invested in this study.

The DCB6Ge tooltip motif, initially described at a Foresight Conference in 2002, was the first complete tooltip ever proposed for diamond mechanosynthesis and remains the only tooltip motif that has been successfully simulated for its intended function on a full 200-atom diamond surface. Although an early paper gives a predicted placement speed of 1 dimer per second for this tooltip, this limit was imposed by the slow speed of recharging the tool using an inefficient recharging method and is not based on any inherent limitation in the speed of use of a charged tooltip. Additionally, no sensing means was proposed for discriminating among the three possible outcomes of an attempted dimer placement—deposition at the correct location, deposition at the wrong location, and failure to place the dimer at all—because the initial proposal was to position the tooltip by dead reckoning, with the proper reaction assured by designing appropriate chemical energetics and relative bond strengths for the tooltip-surface interaction.

More recent theoretical work analyzes a complete set of nine molecular tools made from hydrogen, carbon and germanium able to (a) synthesize all tools in the set (b) recharge all tools in the set from appropriate feedstock molecules and (c) synthesize a wide range of stiff hydrocarbons (diamond, graphite, fullerenes, and the like). All required reactions are analyzed using standard ab initio quantum chemistry methods.

Further research to consider alternate tips will require time-consuming computational chemistry and difficult laboratory work. In the early 2000s, a typical experimental arrangement was to attach a molecule to the tip of an atomic force microscope, and then use the microscope's precise positioning abilities to push the molecule on the tip into another on a substrate. Since the angles and distances can be precisely controlled, and the reaction occurs in a vacuum, novel chemical compounds and arrangements are possible.

The technique of moving single atoms mechanically was proposed by Eric Drexler in his 1986 book The Engines of Creation.

In 1988, researchers at IBM's Zürich Research Institute successfully spelled the letters "IBM" in xenon atoms on a cryogenic copper surface, grossly validating the approach. Since then, a number of research projects have undertaken to use similar techniques to store computer data in a compact fashion. More recently the technique has been used to explore novel physical chemistries, sometimes using lasers to excite the tips to particular energy states, or examine the quantum chemistry of particular chemical bonds.

In 1999, an experimentally proved methodology called feature-oriented scanning (FOS) was suggested. The feature-oriented scanning methodology allows precisely controlling the position of the probe of a scanning probe microscope (SPM) on an atomic surface at room temperature. The suggested methodology supports fully automatic control of single- and multiprobe instruments in solving tasks of mechanosynthesis and bottom-up nanofabrication.

In 2003, Oyabu *et al.* reported the first instance of purely mechanical-based covalent bond-making and bond-breaking, i.e., the first experimental demonstration of true mechanosynthesis—albeit with silicon rather than carbon atoms.

In 2005, the first patent application on diamond mechanosynthesis was filed.

In 2008, a $3.1 million grant was proposed to fund the development of a proof-of-principle mechanosynthesis system.

See also molecular nanotechnology, a more general explanation of the possible products, and discussion of other assembly techniques.

MOLECULAR ENGINEERING

Molecular engineering is an emerging field of study concerned with the design and testing of molecular properties, behavior and interactions in order to assemble better materials, systems, and processes for specific functions. This approach, in which observable properties of a macroscopic system are influenced by direct alteration of a molecular structure, falls into the broader category of "bottom-up" design.

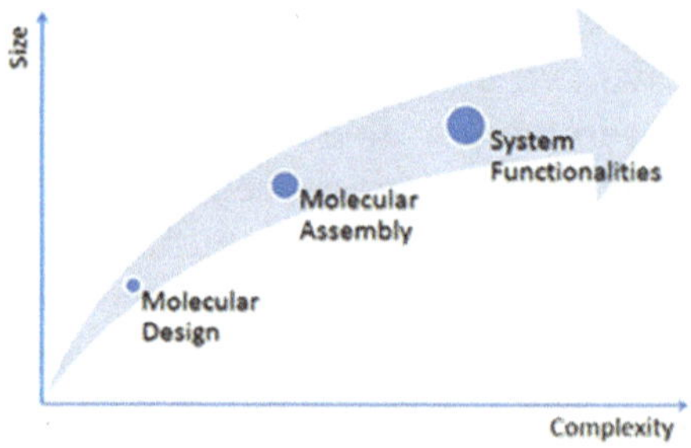

Molecular engineering deals with material development efforts in emerging technologies that require rigorous rational molecular design approaches towards systems of high complexity.

Molecular engineering is highly interdisciplinary by nature, encompassing aspects of chemical engineering, materials science, bioengineering, electrical engineering, physics, mechanical engineering, and chemistry. There is also considerable overlap with nanotechnology, in that both are concerned with the behavior of materials on the scale of nanometers or smaller. Given the highly fundamental nature of molecular interactions, there are a plethora of potential application areas, limited perhaps only by one's imagination and the laws of physics. However, some of the early successes of molecular engineering have come in the fields of immunotherapy, synthetic biology, and printable electronics (see molecular engineering applications).

Molecular engineering is a dynamic and evolving field with complex target problems; breakthroughs require sophisticated and creative engineers who are conversant across disciplines. A rational engineering methodology that is based on molecular principles is in contrast to the widespread trial-and-error approaches common throughout engineering disciplines. Rather than relying on well-described but poorly-understood empirical correlations between the makeup of a system and its properties, a molecular design approach seeks to manipulate system properties directly using an understanding of their chemical and physical origins. This often gives rise to fundamentally new materials and systems, which are required to address outstanding needs in numerous fields, from energy to healthcare to electronics. Additionally, with the increased sophistication of technology, trial-and-error approaches are often costly and difficult, as it may be difficult to account for all relevant dependencies among variables in a complex system. Molecular engineering efforts may include computational tools, experimental methods, or a combination of both.

Molecular engineering was first mentioned in the research literature in 1956 by Arthur R. von Hippel, who defined it as "… a new mode of thinking about engineering problems. Instead of taking prefabricated materials and trying to devise engineering applications consistent with their macroscopic properties, one builds materials from their atoms and molecules for the purpose at hand." This concept was echoed in Richard Feynman's seminal 1959 lecture *There's Plenty of Room at the Bottom*, which is widely regarded as giving birth to some of the fundamental ideas of the field of nanotechnology. In spite of the early introduction of these concepts, it was not until the mid-1980s with the publication of *Engines of Creation: The Coming Era of Nanotechnology* by Drexler that the modern concepts of nano and molecular-scale science began to grow in the public consciousness.

The discovery of electrically-conductive properties in polyacetylene by Alan J. Heeger in 1977 effectively opened the field of organic electronics, which has proved foundational for many molecular engineering efforts. Design and optimization of these materials has led to a number of innovations including organic light-emitting diodes and flexible solar cells.

Applications

Molecular design has been an important element of many disciplines in academia, including bioengineering, chemical engineering, electrical engineering, materials science, mechanical engineering and chemistry. However, one of the on-going challenges is in bringing together the critical mass of manpower amongst disciplines to span the realm from design theory to materials production, and from device design to product development. Thus, while the concept of rational engineering of technology from the bottom-up is not new, it is still far from being widely translated into R&D efforts.

Molecular engineering is used in many industries. Some applications of technologies where molecular engineering plays a critical role:

Consumer Products

- Antibiotic surfaces (eg. incorporation of silver nanoparticles or antibacterial peptides into coatings to prevent microbial infection)
- Cosmetics (eg. rheological modification with small molecules and surfactants in shampoo)
- Cleaning products (eg. nanosilver in laundry detergent)
- Consumer electronics (organic light-emitting diode displays (OLED))
- Electrochromic windows (eg. windows in Dreamliner 787)
- Zero emission vehicles (eg. advanced fuel cells/batteries)
- Self-cleaning surfaces (eg. super hydrophobic surface coatings)

Energy Harvesting and *Storage*

- Flow batteries - Synthesizing molecules for high-energy density electrolytes and highly-selective membranes in grid-scale energy storage systems.
- Lithium-ion batteries - Creating new molecules for use as electrode binders, electrolytes, electrolyte additives, or even for energy storage directly in order to improve energy density (using materials such as graphene, silicon nanorods, and lithium metal), power density, cycle life, and safety.
- Solar cells - Developing new materials for more efficient and cost-effective solar cells including organic, quantum dot or perovskite-based photovoltaics.
- Photocatalytic water splitting - Enhancing the production of hydrogen fuel using solar energy and advanced catalytic materials such as semiconductor nanoparticles

Environmental Engineering

- Water desalination (eg. new membranes for highly-efficient low-cost ion removal)
- Soil remediation (eg. catalytic nanoparticles that accelerate the degradation of long-lived soil contaminants such as chlorinated organic compounds)
- Carbon sequestration (eg. new materials for CO_2 adsorption)

Immunotherapy

- Peptide-based vaccines (e.g. amphiphilic peptide macromolecular assemblies induce a robust immune response)

Synthetic Biology

- CRISPR - Faster and more efficient gene editing technique
- Gene delivery/gene therapy - Designing molecules to deliver modified or new genes into cells of live organisms to cure genetic disorders
- Metabolic engineering - Modifying metabolism of organisms to optimize production of chemicals (eg. synthetic genomics)
- Protein engineering - Altering structure of existing proteins to enable specific new functions, or the creation of fully artificial proteins

Techniques and Instruments Used

Molecular engineers utilize sophisticated tools and instruments to make and analyze the interactions of molecules and the surfaces of materials at the molecular and nano-scale. The complexity of molecules being introduced at the surface is increasing, and the techniques used to analyze surface characteristics at the molecular level are ever-changing and improving. Meantime, advancements in high performance computing have greatly expanded the use of computer simulation in the study of molecular scale systems.

DNA nanotechnology

DNA nanotechnology is the design and manufacture of artificial nucleic acid structures for technological uses. In this field, nucleic acids are used as non-biological engineering materials for nanotechnology rather than as the carriers of genetic information in living cells. Researchers in the field have created static structures such as two- and three-dimensional crystal lattices, nanotubes, polyhedra, and arbitrary shapes, and functional devices such as molecular machines and DNA computers. The field is beginning to be used as a tool to solve basic science problems in structural biology and biophysics, including applications in X-ray crystallography and nuclear magnetic resonance spectroscopy of proteins to determine structures. Potential applications in molecular scale electronics and nanomedicine are also being investigated.

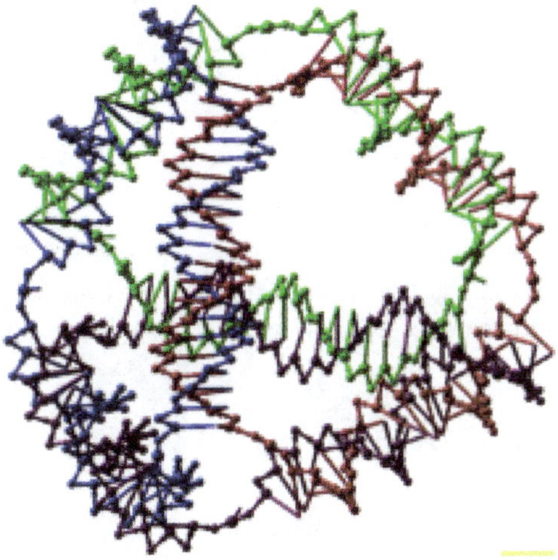

DNA nanotechnology involves forming artificial, designed nanostructures out of nucleic acids, such as this DNA tetrahedron. Each edge of the tetrahedron is a 20 base pair DNA double helix, and each vertex is a three-arm junction. The 4 DNA strands that form the 4 tetrahedral faces are color-coded.

The conceptual foundation for DNA nanotechnology was first laid out by Nadrian Seeman in the early 1980s, and the field began to attract widespread interest in the mid-2000s. This use of nucleic acids is enabled by their strict base pairing rules, which cause only portions of strands with complementary base sequences to bind together to form strong, rigid double helix structures. This allows for the rational design of base sequences that will selectively assemble to form complex target structures with precisely controlled nanoscale features. Several assembly methods are used to make these structures, including tile-based structures that assemble from smaller structures, folding structures using the DNA origami method, and dynamically reconfigurable structures using strand displacement methods. While the field's name specifically references DNA, the same principles have been used with other types of nucleic acids as well, leading to the occasional use of the alternative name *nucleic acid nanotechnology*.

Fundamental Concepts

These four strands associate into a DNA four-arm junction because this structure maximizes the number of correct base pairs, with A matched to T and C matched to G. See this image for a more realistic model of the four-arm junction showing its tertiary structure.

This double-crossover (DX) supramolecular complex consists of five DNA single strands that form two double-helical domains, on the top and the bottom in this image. There are two crossover points where the strands cross from one domain into the other.

Properties of Nucleic Acids

Nanotechnology is often defined as the study of materials and devices with features on a scale below 100 nanometers. DNA nanotechnology, specifically, is an example of bottom-up molecular self-assembly, in which molecular components spontaneously organize into stable structures; the particular form of these structures is induced by the physical and chemical properties of the components selected by the designers. In DNA nanotechnology, the component materials are strands of nucleic acids such as DNA; these strands are often synthetic and are almost always used outside the context of a living cell. DNA is well-suited to nanoscale construction because the binding between two nucleic acid strands depends on simple base pairing rules which are well understood, and form the specific nanoscale structure of the nucleic acid double helix. These qualities make the assembly of nucleic acid structures easy to control through nucleic acid design. This property is absent in other materials used in nanotechnology, including proteins, for which protein design is very difficult, and nanoparticles, which lack the capability for specific assembly on their own.

The structure of a nucleic acid molecule consists of a sequence of nucleotides distinguished by which nucleobase they contain. In DNA, the four bases present are adenine (A), cytosine (C), guanine (G), and thymine (T). Nucleic acids have the property that two molecules will only bind to each other to form a double helix if the two sequences are complementary, meaning that they form matching sequences of base pairs, with A only binding to T, and C only to G. Because the formation of correctly matched base pairs is energetically favorable, nucleic acid strands are expected in most cases to bind to each other in the conformation that maximizes the number of correctly paired bases. The sequences of bases in a system of strands thus determine the pattern of binding and the overall structure in an easily controllable way. In DNA nanotechnology, the base sequences of strands are rationally designed by researchers so that the base pairing interactions cause the strands to assemble in the desired conformation. While DNA is the dominant material used, structures incorporating other nucleic acids such as RNA and peptide nucleic acid (PNA) have also been constructed.

Subfields

DNA nanotechnology is sometimes divided into two overlapping subfields: structural DNA nanotechnology and dynamic DNA nanotechnology. Structural DNA nanotechnology, sometimes abbreviated as SDN, focuses on synthesizing and characterizing nucleic acid complexes and materials that assemble into a static, equilibrium end state. On the other hand, dynamic DNA nanotechnology focuses on complexes with useful non-equilibrium behavior such as the ability to reconfigure based on a chemical or physical stimulus. Some complexes, such as nucleic acid nanomechanical devices, combine features of both the structural and dynamic subfields.

The complexes constructed in structural DNA nanotechnology use topologically branched nucleic acid structures containing junctions. (In contrast, most biological DNA exists as an unbranched double helix.) One of the simplest branched structures is a four-arm junction that consists of four individual DNA strands, portions of which are complementary in a specific pattern. Unlike in natural Holliday junctions, each arm in the artificial immobile four-arm junction has a different base sequence, causing the junction point to be fixed at a certain position. Multiple junctions can be combined in the same complex, such as in the widely used double-crossover (DX) structural motif, which contains two parallel double helical domains with individual strands crossing between the domains at two crossover points. Each crossover point is, topologically, a four-arm junction, but is constrained to one orientation, in contrast to the flexible single four-arm junction, providing a rigidity that makes the DX motif suitable as a structural building block for larger DNA complexes.

Dynamic DNA nanotechnology uses a mechanism called toehold-mediated strand displacement to allow the nucleic acid complexes to reconfigure in response to the addition of a new nucleic acid strand. In this reaction, the incoming strand binds to a single-stranded toehold region of a double-stranded complex, and then displaces one of the strands bound in the original complex through a branch migration process. The overall effect is that one of the strands in the complex is replaced with another one. In addition, reconfigurable structures and devices can be made using functional nucleic acids such as deoxyribozymes and ribozymes, which can perform chemical reactions, and aptamers, which can bind to specific proteins or small molecules.

STRUCTURAL DNA NANOTECHNOLOGY

Structural DNA nanotechnology, sometimes abbreviated as SDN, focuses on synthesizing and characterizing nucleic acid complexes and materials where the assembly has a static, equilibrium endpoint. The nucleic acid double helix has a robust, defined three-dimensional geometry that makes it possible to predict and design the structures of more complicated nucleic acid complexes. Many such structures have been created, including two- and three-dimensional structures, and periodic, aperiodic, and discrete structures.

Extended Lattices

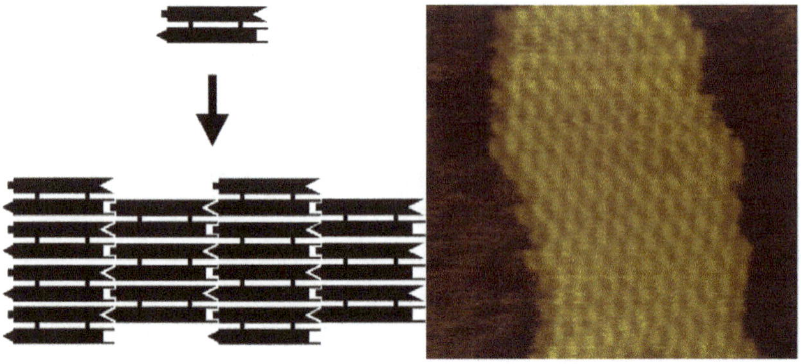

The assembly of a DX array. *Left*, schematic diagram. Each bar represents a double-helical domain of DNA, with the shapes representing complementary sticky ends. The DX complex at top will combine with other DX complexes into the two-dimensional array shown at bottom. *Right*, an atomic force microscopy image of the assembled array. The individual DX tiles are clearly visible within the assembled structure. The field is 150 nm across.

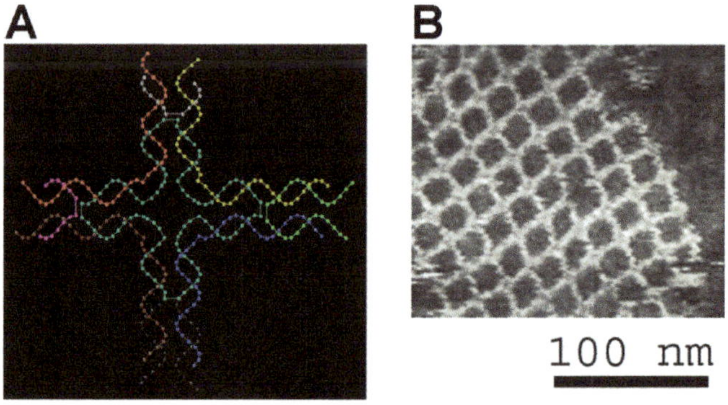

Left, a model of a DNA tile used to make another two-dimensional periodic lattice. *Right*, an atomic force micrograph of the assembled lattice.

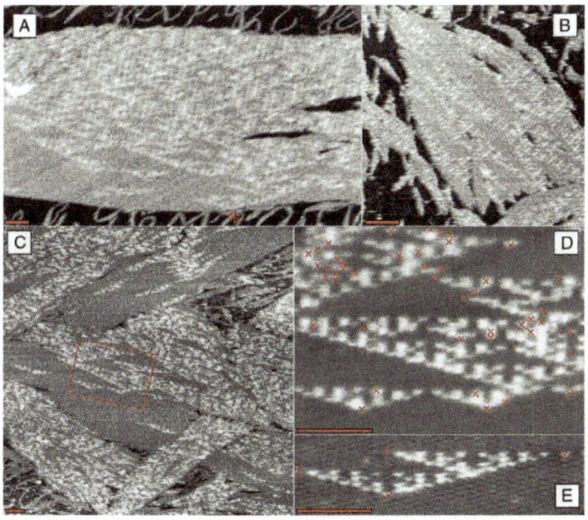

An example of an aperiodic two-dimensional lattice that assembles into a fractal pattern. *Left*, the Sierpinski gasket fractal. *Right*, DNA arrays that display a representation of the Sierpinski gasket on their surfaces

Small nucleic acid complexes can be equipped with sticky ends and combined into larger two-dimensional periodic lattices containing a specific tessellated pattern of the individual molecular tiles. The earliest example of this used double-crossover (DX) complexes as the basic tiles, each containing four sticky ends designed with sequences that caused the DX units to combine into periodic two-dimensional flat sheets that are essentially rigid two-dimensional crystals of DNA.Two-dimensional arrays have been made from other motifs as well, including the Holliday junction rhombus lattice, and various DX-based arrays making use of a double-cohesion scheme. The top two images at right show examples of tile-based periodic lattices.

Two-dimensional arrays can be made to exhibit aperiodic structures whose assembly implements a specific algorithm, exhibiting one form of DNA computing. The DX tiles can have their sticky end sequences chosen so that they act as Wang tiles, allowing them to perform computation. A DX array whose assembly encodes an XOR operation has been demonstrated; this allows the DNA array to implement a cellular automaton that generates a fractal known as the Sierpinski gasket. The third image at right shows this type of array. Another system has the function of a binary counter, displaying a representation of increasing binary numbers as it grows. These results show that computation can be incorporated into the assembly of DNA arrays.

DX arrays have been made to form hollow nanotubes 4–20 nm in diameter, essentially two-dimensional lattices which curve back upon themselves. These DNA nanotubes are somewhat similar in size and shape to carbon nanotubes, and while they lack the electrical conductance of carbon nanotubes, DNA nanotubes are more easily modified and connected to other structures. One of many schemes for constructing DNA nanotubes uses a lattice of curved DX tiles that curls around itself and closes into a tube. In an alternative method that allows the circumference to be specified in a simple, modular fashion using single-stranded tiles, the rigidity of the tube is an emergent property.

Forming three-dimensional lattices of DNA was the earliest goal of DNA nanotechnology, but this proved to be one of the most difficult to realize. Success using a motif based on the concept of tensegrity, a balance between tension and compression forces, was finally reported in 2009.

Discrete Structures

Researchers have synthesized many three-dimensional DNA complexes that each have the connectivity of a polyhedron, such as a cube or octahedron, meaning that the DNA duplexes trace the edges of a polyhedron with a DNA junction at each vertex. The earliest demonstrations of DNA polyhedra were very work-intensive, requiring multiple ligations and solid-phase synthesis steps to create catenated polyhedra. Subsequent work yielded polyhedra whose synthesis was much easier. These include a DNA octahedron made from a long single strand designed to fold into the correct conformation, and a tetrahedron that can be produced from four DNA strands in one step, pictured at the top of this article.

Nanostructures of arbitrary, non-regular shapes are usually made using the DNA origami method. These structures consist of a long, natural virus strand as a "scaffold", which is made to fold into the desired shape by computationally designed short "staple" strands. This method has the advantages of being easy to design, as the base sequence is predetermined by the scaffold strand sequence, and not requiring high strand purity and accurate stoichiometry, as most other DNA nanotechnology methods do. DNA origami was first demonstrated for two-dimensional shapes, such as a smiley face and a coarse map of the Western Hemisphere. Solid three-dimensional structures can be made by using parallel DNA helices arranged in a honeycomb pattern, and structures with two-dimensional faces can be made to fold into a hollow overall three-dimensional shape, akin to a cardboard box. These can be programmed to open and reveal or release a molecular cargo in response to a stimulus, making them potentially useful as programmable molecular cages.

Templated Assembly

Nucleic acid structures can be made to incorporate molecules other than nucleic acids, sometimes called heteroelements, including proteins, metallic nanoparticles, quantum dots, and fullerenes. This allows the construction of materials and devices with a range of functionalities much greater than is possible with nucleic acids alone. The goal is to use the self-assembly of the nucleic acid structures to template the assembly of the nanoparticles hosted on them, controlling their position and in some cases orientation. Many of these schemes use a covalent attachment scheme, using oligonucleotides with amide or thiol functional groups as a chemical handle to bind the heteroelements. This covalent binding scheme has been used to arrange gold nanoparticles on a DX-based array, and to arrange streptavidin protein molecules into specific patterns on a DX array. A non-covalent hosting scheme using Dervan polyamides on a DX array was used to arrange streptavidin proteins in a specific pattern on a DX array. Carbon nanotubes have been hosted on DNA arrays in a pattern allowing the assembly to act as a molecular electronic device, a carbon nanotube field-effect transistor. In addition, there are nucleic acid metallization methods, in which the nucleic acid is replaced by a metal which assumes the general shape of the original nucleic acid structure, and schemes for using nucleic acid nanostructures as lithography masks, transferring their pattern into a solid surface.

DYNAMIC DNA NANOTECHNOLOGY

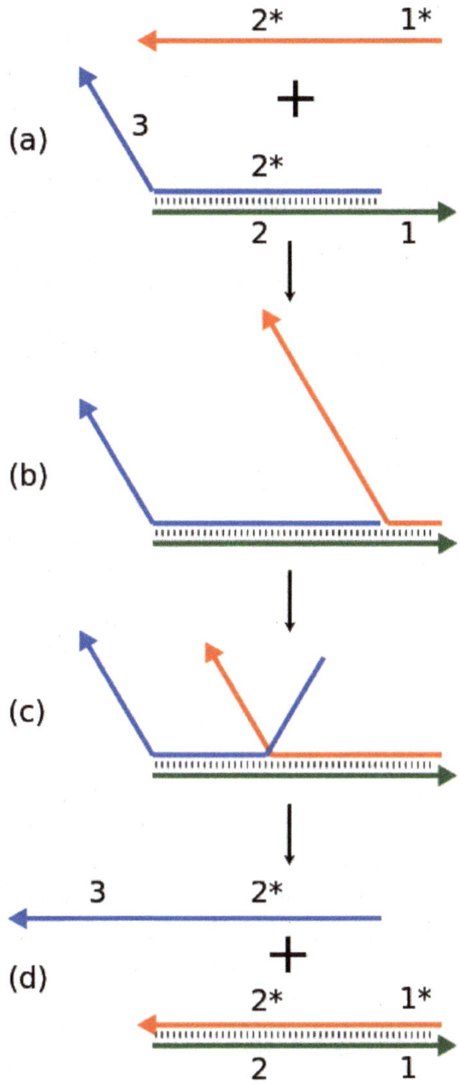

Dynamic DNA nanotechnology often makes use of toehold-mediated strand displacement reactions. In this example, the red strand binds to the single stranded toehold region on the green strand (region 1), and then in a branch migration process across region 2, the blue strand is displaced and freed from the complex. Reactions like these are used to dynamically reconfigure or assemble nucleic acid nanostructures. In addition, the red and blue strands can be used as signals in a molecular logic gate.

Dynamic DNA nanotechnology focuses on forming nucleic acid systems with designed dynamic functionalities related to their overall structures, such as computation and mechanical motion. There is some overlap between structural and dynamic DNA nanotechnology, as structures can be formed through annealing and then reconfigured dynamically, or can be made to form dynamically in the first place.

Nanomechanical Devices

DNA complexes have been made that change their conformation upon some stimulus, making them one form of nanorobotics. These structures are initially formed in the same way as the static structures made in structural DNA nanotechnology, but are designed so that dynamic reconfiguration is possible after the initial assembly.The earliest such device made use of the transition between the B-DNA and Z-DNA forms to respond to a change in buffer conditions by undergoing a twisting motion.This reliance on buffer conditions, however, caused all devices to change state at the same time. Subsequent systems could change states based upon the presence of control strands, allowing multiple devices to be independently operated in solution. Some examples of such systems are a "molecular tweezers" design that has an open and a closed state, a device that could switch from a paranemic-crossover (PX) conformation to a double-junction (JX2) conformation, undergoing rotational motion in the process, and a two-dimensional array that could dynamically expand and contract in response to control strands. Structures have also been made that dynamically open or close, potentially acting as a molecular cage to release or reveal a functional cargo upon opening.

DNA walkers are a class of nucleic acid nanomachines that exhibit directional motion along a linear track. A large number of schemes have been demonstrated. One strategy is to control the motion of the walker along the track using control strands that need to be manually added in sequence. Another approach is to make use of restriction enzymes or deoxyribozymes to cleave the strands and cause the walker to move forward, which has the advantage of running autonomously. A later system could walk upon a two-dimensional surface rather than a linear track, and demonstrated the ability to selectively pick up and move molecular cargo.Additionally, a linear walker has been demonstrated that performs DNA-templated synthesis as the walker advances along the track, allowing autonomous multistep chemical synthesis directed by the walker. The synthetic DNA walkers' function is similar to that of the proteins dynein and kinesin.

Strand Displacement Cascades

Cascades of strand displacement reactions can be used for either computational or structural purposes. An individual strand displacement reaction involves revealing a new sequence in response to the presence of some initiator strand. Many such reactions can be linked into a cascade where the newly revealed output sequence of one reaction can initiate another strand displacement reaction elsewhere. This in turn allows for the construction of chemical reaction networks with many components, exhibiting complex computational and information processing abilities. These cascades are made energetically favorable through the formation of new base pairs, and the entropy gain from disassembly reactions. Strand displacement cascades allow isothermal operation of the assembly or computational process, in contrast to traditional nucleic acid assembly's requirement for a thermal annealing step, where the temperature is raised and then slowly lowered to ensure proper formation of the desired structure. They can also support catalytic function of the initiator species, where less than one equivalent of the initiator can cause the reaction to go to completion.

Strand displacement complexes can be used to make molecular logic gates capable of complex computation. Unlike traditional electronic computers, which use electric current as inputs and outputs, molecular computers use the concentrations of specific chemical species as signals. In the case of nucleic acid strand displacement circuits, the signal is the presence of nucleic acid strands that are released or consumed by binding and unbinding events to other strands in displacement complexes. This approach has been used to make logic gates such as AND, OR, and NOT gates. More recently, a four-bit circuit was demonstrated that can compute the square root of the integers 0–15, using a system of gates containing 130 DNA strands.

Another use of strand displacement cascades is to make dynamically assembled structures. These use a hairpin structure for the reactants, so that when the input strand binds, the newly revealed sequence is on the same molecule rather than disassembling. This allows new opened hairpins to be added to a growing complex. This approach has been used to make simple structures such as three- and four-arm junctions and dendrimers.

APPLICATIONS

DNA nanotechnology provides one of the few ways to form designed, complex structures with precise control over nanoscale features. The field is beginning to see application to solve basic science problems in structural biology and biophysics. The earliest such application envisaged for the field, and one still in development, is in crystallography, where molecules that are difficult to crystallize in isolation could be arranged within a three-dimensional nucleic acid lattice, allowing determination of their structure. Another application is the use of DNA origami rods to replace liquid crystals in residual dipolar coupling experiments in protein NMR spectroscopy; using DNA origami is advantageous because, unlike liquid crystals, they are tolerant of the detergents needed to suspend membrane proteins in solution. DNA walkers have been used as nanoscale assembly lines to move nanoparticles and direct chemical synthesis. Further, DNA origami structures have aided in the biophysical studies of enzyme function and protein folding.

DNA nanotechnology is moving toward potential real-world applications. The ability of nucleic acid arrays to arrange other molecules indicates its potential applications in molecular scale electronics. The assembly of a nucleic acid structure could be used to template the assembly of a molecular electronic elements such as molecular wires, providing a method for nanometer-scale control of the placement and overall architecture of the device analogous to a molecular breadboard. DNA nanotechnology has been compared to the concept of programmable matter because of the coupling of computation to its material properties.

In a study conducted by a group of scientists from iNANO center and CDNA Center in Aarhus university (Aarhus), researchers were able to construct a small multi-switchable

3D DNA Box Origami. The proposed nanoparticle was characterized by atomic force microscopy (AFM), transmission electron microscopy (TEM) and Förster resonance energy transfer (FRET). The constructed box was shown to have a unique reclosing mechanism, which enabled it to repeatedly open and close in response to a unique set of DNA or RNA keys. The authors proposed that this "DNA device can potentially be used for a broad range of applications such as controlling the function of single molecules, controlled drug delivery, and molecular computing.".

There are potential applications for DNA nanotechnology in nanomedicine, making use of its ability to perform computation in a biocompatible format to make "smart drugs" for targeted drug delivery. One such system being investigated uses a hollow DNA box containing proteins that induce apoptosis, or cell death, that will only open when in proximity to a cancer cell.There has additionally been interest in expressing these artificial structures in engineered living bacterial cells, most likely using the transcribed RNA for the assembly, although it is unknown whether these complex structures are able to efficiently fold or assemble in the cell's cytoplasm. If successful, this could enable directed evolution of nucleic acid nanostructures.Scientists at Oxford University reported the self-assembly of four short strands of synthetic DNA into a cage which can enter cells and survive for at least 48 hours. The fluorescently labeled DNA tetrahedra were found to remain intact in the laboratory cultured human kidney cells despite the attack by cellular enzymes after two days. This experiment showed the potential of drug delivery inside the living cells using the DNA 'cage'. A DNA tetrahedron was used to deliver RNA Interference (RNAi) in a mouse model, reported a team of researchers in MIT. Delivery of the interfering RNA for treatment has showed some success using polymer or lipid, but there are limits of safety and imprecise targeting, in addition to short shelf life in the blood stream. The DNA nanostructure created by the team consists of six strands of DNA to form a tetrahedron, with one strand of RNA affixed to each of the six edges. The tetrahedron is further equipped with targeting protein, three folate molecules, which lead the DNA nanoparticles to the abundant folate receptors found on some tumors. The result showed that the gene expression targeted by the RNAi, luciferase, dropped by more than half. This study shows promise in using DNA nanotechnology as an effective tool to deliver treatment using the emerging RNA Interference technology.

DESIGN

DNA nanostructures must be rationally designed so that individual nucleic acid strands will assemble into the desired structures. This process usually begins with specification of a desired target structure or function. Then, the overall secondary structure of the target complex is determined, specifying the arrangement of nucleic acid strands within the structure, and which portions of those strands should be bound to each other. The last step is the primary structure design, which is the specification of the actual base sequences of each nucleic acid strand.

Structural Design

The first step in designing a nucleic acid nanostructure is to decide how a given structure should be represented by a specific arrangement of nucleic acid strands. This design step determines the secondary structure, or the positions of the base pairs that hold the individual strands together in the desired shape. Several approaches have been demonstrated:

- **Tile-based structures.** This approach breaks the target structure into smaller units with strong binding between the strands contained in each unit, and weaker interactions between the units. It is often used to make periodic lattices, but can also be used to implement algorithmic self-assembly, making them a platform for DNA computing. This was the dominant design strategy used from the mid-1990s until the mid-2000s, when the DNA origami methodology was developed.
- **Folding structures.** An alternative to the tile-based approach, folding approaches make the nanostructure from one long strand, which can either have a designed sequence that folds due to its interactions with itself, or it can be folded into the desired shape by using shorter, "staple" strands. This latter method is called DNA origami, which allows forming nanoscale two- and three-dimensional shapes (see Discrete structures above).
- **Dynamic assembly.** This approach directly controls the kinetics of DNA self-assembly, specifying all of the intermediate steps in the reaction mechanism in addition to the final product. This is done using starting materials which adopt a hairpin structure; these then assemble into the final conformation in a cascade reaction, in a specific order (see Strand displacement cascades below). This approach has the advantage of proceeding isothermally, at a constant temperature. This is in contrast to the thermodynamic approaches, which require a thermal annealing step where a temperature change is required to trigger the assembly and favor proper formation of the desired structure.

Sequence Design

After any of the above approaches are used to design the secondary structure of a target complex, an actual sequence of nucleotides that will form into the desired structure must be devised. Nucleic acid design is the process of assigning a specific nucleic acid base sequence to each of a structure's constituent strands so that they will associate into a desired conformation. Most methods have the goal of designing sequences so that the target structure has the lowest energy, and is thus the most thermodynamically favorable, while incorrectly assembled structures have higher energies and are thus disfavored. This is done either through simple, faster heuristic methods such as sequence symmetry minimization, or by using a full nearest-neighbor thermodynamic model, which is more accurate but slower and more computationally intensive. Geometric models are used to examine tertiary structure of the nanostructures and to ensure that the complexes are not overly strained.

Nucleic acid design has similar goals to protein design. In both, the sequence of monomers is designed to favor the desired target structure and to disfavor other structures.

Nucleic acid design has the advantage of being much computationally easier than protein design, because the simple base pairing rules are sufficient to predict a structure's energetic favorability, and detailed information about the overall three-dimensional folding of the structure is not required. This allows the use of simple heuristic methods that yield experimentally robust designs. However, nucleic acid structures are less versatile than proteins in their function because of proteins' increased ability to fold into complex structures, and the limited chemical diversity of the four nucleotides as compared to the twenty proteinogenic amino acids.

MATERIALS AND METHODS

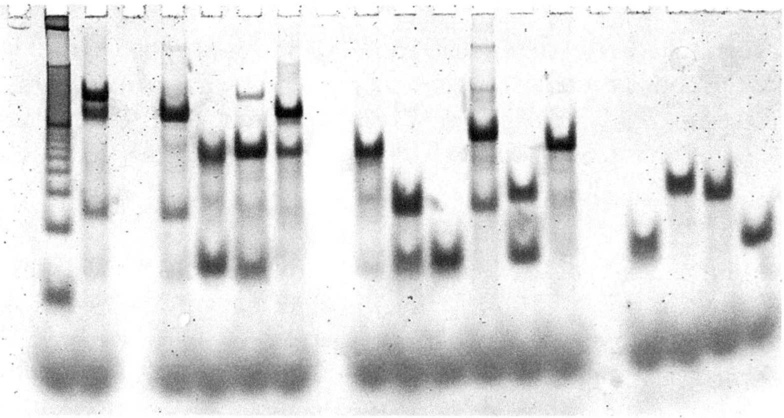

Gel electrophoresis methods, such as this formation assay on a DX complex, are used to ascertain whether the desired structures are forming properly. Each vertical lane contains a series of bands, where each band is characteristic of a particular reaction intermediate.

The sequences of the DNA strands making up a target structure are designed computationally, using molecular modeling and thermodynamic modeling software. The nucleic acids themselves are then synthesized using standard oligonucleotide synthesis methods, usually automated in an oligonucleotide synthesizer, and strands of custom sequences are commercially available. Strands can be purified by denaturing gel electrophoresis if needed, and precise concentrations determined via any of several nucleic acid quantitation methods using ultraviolet absorbance spectroscopy.

The fully formed target structures can be verified using native gel electrophoresis, which gives size and shape information for the nucleic acid complexes. An electrophoretic mobility shift assay can assess whether a structure incorporates all desired strands. Fluorescent labeling and Förster resonance energy transfer (FRET) are sometimes used to characterize the structure of the complexes.

Nucleic acid structures can be directly imaged by atomic force microscopy, which is well suited to extended two-dimensional structures, but less useful for discrete three-dimensional structures because of the microscope tip's interaction with the fragile nucleic acid structure; transmission electron microscopy and cryo-electron microscopy are often used in this case. Extended three-dimensional lattices are analyzed by X-ray crystallography.

HISTORY

The woodcut *Depth* (pictured) by M. C. Escher reportedly inspired Nadrian Seeman to consider using three-dimensional lattices of DNA to orient hard-to-crystallize molecules. This led to the beginning of the field of DNA nanotechnology.

The conceptual foundation for DNA nanotechnology was first laid out by Nadrian Seeman in the early 1980s. Seeman's original motivation was to create a three-dimensional DNA lattice for orienting other large molecules, which would simplify their crystallographic study by eliminating the difficult process of obtaining pure crystals. This idea had reportedly come to him in late 1980, after realizing the similarity between the woodcut *Depth* by M. C. Escher and an array of DNA six-arm junctions. Several natural branched DNA structures were known at the time, including the DNA replication fork and the mobile Holliday junction, but Seeman's insight was that immobile nucleic acid junctions could be created by properly designing the strand sequences to remove symmetry in the assembled molecule, and that these immobile junctions could in principle be combined into rigid crystalline lattices. The first theoretical paper proposing this scheme was published in 1982, and the first experimental demonstration of an immobile DNA junction was published the following year.

In 1991, Seeman's laboratory published a report on the synthesis of a cube made of DNA, the first synthetic three-dimensional nucleic acid nanostructure, for which he received the 1995 Feynman Prize in Nanotechnology. This was followed by a DNA truncated octahedron. However, it soon became clear that these structures, polygonal shapes with flexible junctions as their vertices, were not rigid enough to form extended three-dimensional lattices. Seeman developed the more rigid double-crossover (DX) structural motif, and in 1998, in collaboration with Erik Winfree, published the creation of two-dimensional lattices of DX tiles. These tile-based structures had the advantage that they provided the ability to implement DNA computing, which was demonstrated by Winfree and Paul Rothemund in their 2004 paper on the algorithmic self-assembly of a Sierpinski gasket structure, and for which they shared the 2006 Feynman Prize in Nanotechnology. Winfree's key insight was that the DX tiles could be used as Wang tiles, meaning that their assembly could perform computation. The synthesis of a three-dimensional lattice was finally published by Seeman in 2009, nearly thirty years after he had set out to achieve it.

New abilities continued to be discovered for designed DNA structures throughout the 2000s. The first DNA nanomachine—a motif that changes its structure in response to an input—was demonstrated in 1999 by Seeman. An improved system, which was the first nucleic acid device to make use of toehold-mediated strand displacement, was demonstrated by Bernard Yurke the following year. The next advance was to translate this into mechanical motion, and in 2004 and 2005, several DNA walker systems were demonstrated by the groups of Seeman, Niles Pierce, Andrew Turberfield, and Chengde Mao. The idea of using DNA arrays to template the assembly of other molecules such as nanoparticles and proteins, first suggested by Bruche Robinson and Seeman in 1987, was demonstrated in 2002 by Seeman, Kiehl et al. and subsequently by many other groups.

In 2006, Rothemund first demonstrated the DNA origami method for easily and robustly forming folded DNA structures of arbitrary shape. Rothemund had conceived of this method as being conceptually intermediate between Seeman's DX lattices, which used many short strands, and William Shih's DNA octahedron, which consisted mostly of one very long strand. Rothemund's DNA origami contains a long strand which folding is assisted by several short strands. This method allowed forming much larger structures than formerly possible, and which are less technically demanding to design and synthesize. DNA origami was the cover story of *Nature* on March 15, 2006. Rothemund's research demonstrating two-dimensional DNA origami structures was followed by the demonstration of solid three-dimensional DNA origami by Douglas *et al.* in 2009,while the labs of Jørgen Kjems and Yan demonstrated hollow three-dimensional structures made out of two-dimensional faces.

DNA nanotechnology was initially met with some skepticism due to the unusual non-biological use of nucleic acids as materials for building structures and doing computation, and the preponderance of proof of principle experiments that extended the abilities of the field but were far from actual applications. Seeman's 1991 paper on the synthesis of the DNA cube was rejected by the journal *Science* after one reviewer praised its originality while another criticized it for its lack of biological relevance. By the early 2010s, however, the field was considered to have increased its abilities to the point that applications for basic science research were beginning to be realized, and practical applications in medicine and other fields were beginning to be considered feasible. The field had grown from very few active laboratories in 2001 to at least 60 in 2010, which increased the talent pool and thus the number of scientific advances in the field during that decade.

HEART NANOTECHNOLOGY

Heart nanotechnology is the "Engineering of functional systems at the molecular scale" ("Nanotechnology Research").

Nanotechnology

Nanotechnology deals with structures and materials that are approximately one to one-hundred nanometers in length. At this microscopic level, quantum mechanics take place and are in effect, resulting in behaviors that would seem quite strange compared to what humans see with the naked eye (regular matter). Nanotechnology is used for a wide variety of fields of technology, ranging from energy to electronics to medicine. In the category of medicine, nanotechnology is still relatively new and has not yet been widely adopted by the field. It's possible that nanotechnology could be the new breakthrough of medicine and may eventually be the solution and cure for many of the health problems that humans encounter. Nanotechnology may lead to the cure for illnesses such as the common cold, diseases, and cancer. It is already starting to be used as a treatment for some serious health issues; more specifically it is being used to treat the heart and cancer.

Nanomedicine

Nanotechnology in the field of medicine is more commonly referred to as nanomedicine. Nanomedicine that deals with helping the heart is really starting to take off and gain in popularity compared to most of the other fields that nanomedicine currently has to offer. There are several heart problems that nanotechnology has promising evidence of being effective in the treatment of heart disease in the near future.

Examples

It should hopefully be able to treat heart valves that are defective; and detect and treat arterial plaque in the heart ("Nanotechnology Made Clear"). Nanomedicine should be able to help heal the hearts of people that have already been victims of heart disease and heart attacks. On the other hand, it will also play a key role in finding people with a high risk of having heart disease, and will be able to help prevent heart attacks from happening in the first place. Nanotechnology of the heart is a lot less invasive than surgery because everything is occurring at a minuscule level in the body compared to relatively large tissues that are dealt with in surgery. With our technology today, heart surgeries are performed to treat the damaged heart tissue that resulted from a heart attack. This is a major surgery that usually takes a couple of months to recover from ("WebMD - Better Information. Better Health"). During this period, patients are extremely limited in the activities that they can do. This long recovery process is an inconvenience to the patients, and with the growth of medicine it most likely won't be very long before a more efficient method for treating heart attack patients will be developed and used. The method that is the frontrunner to replace major heart surgery is the use of nanotechnology. There are a couple alternate ways to heart surgery that nanotechnology will potentially be able to offer in the future.

Alternatives to Surgery

With people that have heart disease or that have suffered a heart attack, their hearts are often damaged and weakened. The more minor forms of heart failure do not require surgery and are often treated with medications ("WebMD - Better Information. Better Health"). The use of nanotechnology on treating damaged hearts will not replace these milder heart problems, but rather the more serious heart problems that currently require surgery or sometimes even heart transplants.

Heart Repair

A group of engineers, doctors and materials scientists at MIT and Children's Hospital Boston have teamed together and are starting the movement of finding a way to use nanotechnology to strengthen the weakened heart tissue ("MIT - Massachusetts Institute of Technology"). The first method uses nanotechnology combined with tissue engineering, and gold nanowires are placed and woven into the damaged parts of the heart, essentially replacing the non-functioning or dead tissues.

Tissue Regeneration

The other approach would potentially use minuscule nanoparticles that would travel through the body and find dying heart tissue. The nanoparticles would be carrying objects such as "stem cells, growth factors, drugs and other therapeutic compounds,". Then the nanoparticles would release the compounds and inject them into the damaged heart tissue. This would theoretically lead to the regeneration of the tissue.

Heart repair Difficulties

Being able to fix cardiac tissue that has been damaged from a heart attack or heart disease is not very simple and it is one of the major challenges today in the field of tissue engineering ("Popular Science"). This is because heart cells are not the easiest objects to create in a lab. It takes an enormous amount of special care and work to develop the cells so that they beat in sync with one another ("Popular Science"). Even after the heart cells have finally been made, it is also a large task to insert the cells into the inoperable parts of the heart and to get them working in unison with the tissues that were still working properly ("Popular Science").

Heart Patches

There have been several successful examples of this with the use of a "stem-cell- based heart patch developed by Duke University researchers," ("Popular Science"). The biomaterials that make up the patch are usually made of either biological polymers like alginate or synthetic polymers such as polylactic acid ("Nature Nanotechnology"). These materials are good at organizing the cells into functioning tissues; however they act as insulators and are poor conductors of electricity, which is a major problem especially in the heart ("Nature Nanotechnology"). Since the electrical signals that are sent between calcium ions are what control when the cardiomyocytes of the heart contract, which makes the heart beat, the stem-cell heart patch is not very efficient and not as effective as doctors would like it to be ("Popular Science"). The results of the patch not being very conductive is that the cells are not able to attain a smooth, continuous beat throughout the entire tissue containing the stem cells. This results in the heart not functioning properly, which in turn could mean that more heart problems might arise due to the implanting of the stem cells.

Tissue Scaffolds

Recently there have been some new developments in the field of nanotechnology that will be more efficient than the poorly conducting stem-cell-based patch ("Nature Nanotechnology"). Scientists and researchers found a way for these stem cell patches (also known as tissue scaffolds) to be conductive and therefore become exponentially more effective ("Nature Nanotechnology"). They found that by growing gold nanowires into and through the patches, they were able to greatly increase the electrical conductivity. The nanowires are thicker than the original scaffold and the cells are better organized as well. There is also an increase in production of the proteins needed for muscle calcium binding and contraction. The gold nanowires poke through the stem cell's scaffolding material, which strengthens the electrical communication between surrounding heart cells. Without the nanowires, the stem cell patches produced a minute current and the cells would only beat in small clusters at the stimulation origin. With the nanowires, the cells seem to contract together even when they are clustered far away from the source of stimulation. The use of gold nanowires with the stem cell heart patches is still a relatively new concept and it will probably be awhile before they will be used in humans. It is hoped that the nanowires will be tested in live animals in the near future.

Nanoparticles

Another way that nanotechnology will potentially be used to help fix damaged heart tissues is through the use of guided nanoparticle "missiles". These nanoparticles can cling to and attach to artery walls and secrete medicine at a slow rate ("MIT-Massachusetts Institute of Technology"). The particles, known as nanoburrs due to the fact that they are coated with little protein fragments that stick to and target certain proteins. The nanoburrs can be made to release the drug that is attached to them over the course of several days ("MIT-Massachusetts Institute of Technology"). They are unique compared to regular drugs because they can find the particular damaged tissue, attach to it, and release the drug payload that is attached to it ("MIT-Massachusetts Institute of Technology"). What happens is the nanoburrs are targeted to a certain structure, known as the basement membrane; this membrane lines the arterial walls and is only present if the area is damaged. Nanoburrs could be able to carry drugs that are effective in treating the heart, and also potentially carry stem cells to help regenerate the damaged heart tissue ("MIT-Massachusetts Institute of Technology").

Composition

The particles are made up of three different layers and are sixty nanometers in diameter ("MIT-Massachusetts Institute of Technology").The outer layer is a coating of polymer called PEG, and its job is to protect the drug from disintegrating while it is traveling through the body. The middle layer consists of a fatty substance and the inner core contains the actual drug along with a polymer chain, which controls the amount of time it will take before the drug is released ("MIT-Massachusetts Institute of Technology").

Research

In a study done on rats, the nanoparticles were injected directly into the rat's tail and they still were able to reach the desired target (the left carotid artery) at a rate that was twice the amount of the non-targeted nanoparticles ("MIT-Massachusetts Institute of Technology"). Because the particles can deliver drugs over a long period of time, and can be injected intravenously, the patients would not need to have multiple repeated injections, or invasive surgeries on the heart which would be a lot more convenient. The only downside to this is that the existing delivery approaches are invasive, requiring either a direct injection into the heart, catheter procedures, or surgical implants. There is no question, however, that the future of heart repairs and heart disease/attack prevention will definitely involve the use of nanotechnology in some way.

Polyketal Nanoparticles

Composition

Polyketal nanoparticles are pH-sensitive, hydrophobic nanoparticles formulated from poly(1-4-phenyleneacetone dimethylene ketal). They are an acid-sensitive vehicle of drug delivery, specifically designed for targeting the environments of tumors, phagosomes, and inflammatory tissue. In such acidic environments, these nanoparticles undergo accelerated hydrolysis into low molecular weight hydrophilic compounds, consequently releasing their therapeutic contents at a faster rate. Unlike polyester-based nanoparticles, polyketal nanoparticles do not generate acidic degradation products following hydrolysis

Use in Myocardial Infarction

Post-myocardial infarction, inflammatory leukocytes invade the myocardium. Leukocytes contain high amounts of Nicotinamide adenine dinucleotide phosphate (NADPH) and Nox2. Nox2 and NADPH oxidase combine to act as a major source of cardiac superoxide production, which in excess can lead to myocyte hypertrophy, apoptosis, fibrosis, and increased matrix metalloproteinase-2 expression. In a mouse-model study by Somasuntharam et al. 2013, polyketal nanoparticles were used as a delivery vehicle for siRNA to target and inhibit Nox2 in the infarcted heart. Following intramyocardial injection in vivo, Nox2-siRNA nanoparticles prevented upregulation of Nox2-NADPH oxidase, and improved fractional shortening. When taken up by macrophages in the myocardium following a MI, the nanoparticles degraded in the acidic environment of the endosomes/phagosomes, releasing Nox2-specific siRNA into the cytoplasm.

Polyketal nanoparticles have also been used in the infarcted mouse heart to prevent ischemia-reperfusion injury caused by reactive oxygen species (ROS). Levels of the antioxidant Cu/Zn-superoxide dismutase (SOD1), which scavenges harmful ROS, decrease following MI. SOD1-enacapsulated polyketal nanoparticles are able to scavenge reperfusion-injury induced ROS. Furthermore, this treatment improved fractional shortening, suggesting the benefit of targeted delivery by polyketals. One of the key advantages of polyketal use is that they do not exacerbate the inflammatory response, even when administered at concentrations exceeding therapeutic limits. In contrast to commonly used poly(lactic-co-glycolic acid) (PLGA) nanoparticles, polyketal nanoparticle administration in mice instigates little recruitment of inflammatory cells. Additionally, intramuscular injection of polyketals into the leg of rats shows no significant increases in inflammatory cytokines such as IL-6, IL-1ß, TNF-α and IL-12.

NANOMEDICINE

Nanomedicine is the medical application of nanotechnology. Nanomedicine ranges from the medical applications of nanomaterials and biological devices, to nanoelectronic biosensors, and even possible future applications of molecular nanotechnology such as biological machines. Current problems for nanomedicine involve understanding the issues related to toxicity and environmental impact of nanoscale materials (materials whose structure is on the scale of nanometers, i.e. billionths of a meter).

Functionalities can be added to nanomaterials by interfacing them with biological molecules or structures. The size of nanomaterials is similar to that of most biological molecules and structures; therefore, nanomaterials can be useful for both in vivo and in vitro biomedical research and applications. Thus far, the integration of nanomaterials with biology has led to the development of diagnostic devices, contrast agents, analytical tools, physical therapy applications, and drug delivery vehicles.

Nanomedicine seeks to deliver a valuable set of research tools and clinically useful devices in the near future. The National Nanotechnology Initiative expects new commercial applications in the pharmaceutical industry that may include advanced drug delivery systems, new therapies, and in vivo imaging. Nanomedicine research is receiving funding from the US National Institutes of Health, including the funding in 2005 of a five-year plan to set up four nanomedicine centers.

Nanomedicine sales reached $16 billion in 2015, with a minimum of $3.8 billion in nanotechnology R&D being invested every year. Global funding for emerging nanotechnology increased by 45% per year in recent years, with product sales exceeding $1 trillion in 2013. As the nanomedicine industry continues to grow, it is expected to have a significant impact on the economy.

Drug Delivery

Nanotechnology has provided the possibility of delivering drugs to specific cells using nanoparticles.

The overall drug consumption and side-effects may be lowered significantly by depositing the active agent in the morbid region only and in no higher dose than needed. Targeted drug delivery is intended to reduce the side effects of drugs with concomitant decreases in consumption and treatment expenses. Drug delivery focuses on maximizing bioavailability both at specific places in the body and over a period of time. This can potentially be achieved by molecular targeting by nanoengineered devices. More than $65 billion are wasted each year due to poor bioavailability. A benefit of using nanoscale for medical technologies is that smaller devices are less invasive and can possibly be implanted inside the body, plus biochemical reaction times are much shorter. These devices are faster and more sensitive than typical drug delivery. The efficacy of drug delivery through nanomedicine is largely based upon: a) efficient encapsulation of the drugs, b) successful delivery of drug to the targeted region of the body, and c) successful release of the drug.

Drug delivery systems, lipid- or polymer-based nanoparticles, can be designed to improve the pharmacokinetics and biodistribution of the drug. However, the pharmacokinetics and pharmacodynamics of nanomedicine is highly variable among different patients. When designed to avoid the body's defence mechanisms,nanoparticles have beneficial properties that can be used to improve drug delivery. Complex drug delivery mechanisms are being developed, including the ability to get drugs through cell membranes and into cell cytoplasm. Triggered response is one way for drug molecules to be used more efficiently. Drugs are placed in the body and only activate on encountering a particular signal. For example, a drug with poor solubility will be replaced by a drug delivery system where both hydrophilic and hydrophobic environments exist, improving the solubility. Drug delivery systems may also be able to prevent tissue damage through regulated drug release; reduce drug clearance rates; or lower the volume of distribution and reduce the effect on non-target tissue. However, the biodistribution of these nanoparticles is still imperfect due to the complex host's reactions to nano- and microsized materials and the difficulty in targeting specific organs in the body. Nevertheless, a lot of work is still ongoing to optimize and better understand the potential and limitations of nanoparticulate systems. While advancement of research proves that targeting and distribution can be augmented by nanoparticles, the dangers of nanotoxicity become an important next step in further understanding of their medical uses.

Nanoparticles can be used in combination therapy for decreasing antibiotic resistance or for their antimicrobial properties. Nanoparticles might also used to circumvent multidrug resistance (MDR) mechanisms.

Types of systems used

Two forms of nanomedicine that have already been tested in mice and are awaiting human trials that will be using gold nanoshells to help diagnose and treat cancer, and using liposomes as vaccine adjuvants and as vehicles for drug transport. Similarly, drug detoxification is also another application for nanomedicine which has shown promising results in rats. Advances in Lipid nanotechnology was also instrumental in engineering medical nanodevices and novel drug delivery systems as well as in developing sensing applications. Another example can be found in dendrimers and nanoporous materials. Another example is to use block co-polymers, which form micelles for drug encapsulation.

Polymeric nano-particles are a competing technology to lipidic (based mainly on Phospholipids) nano-particles. There is an additional risk of toxicity associated with polymers not widely studied or understood. The major advantages of polymers is stability, lower cost and predictable characterisation. However, in the patient's body this very stability (slow degradation) is a negative factor. Phospholipids on the other hand are membrane lipids (already present in the body and surrounding each cell), have a GRAS (Generally Recognised As Safe) status from FDA and are derived from natural sources without any complex chemistry involved. They are not metabolised but rather absorbed by the body and the degradation products are themselves nutrients (fats or micronutrients).

Protein and peptides exert multiple biological actions in the human body and they have been identified as showing great promise for treatment of various diseases and disorders. These macromolecules are called biopharmaceuticals. Targeted and/or controlled delivery of these biopharmaceuticals using nanomaterials like nanoparticles and Dendrimers is an emerging field called nanobiopharmaceutics, and these products are called nanobiopharmaceuticals.

Another highly efficient system for microRNA delivery for example are nanoparticles formed by the self-assembly of two different microRNAs deregulated in cancer.

Another vision is based on small electromechanical systems; nanoelectromechanical systems are being investigated for the active release of drugs. Some potentially important applications include cancer treatment with iron nanoparticles or gold shells.Nanotechnology is also opening up new opportunities in implantable delivery systems, which are often preferable to the use of injectable drugs, because the latter frequently display first-order kinetics (the blood concentration goes up rapidly, but drops exponentially over time). This rapid rise may cause difficulties with toxicity, and drug efficacy can diminish as the drug concentration falls below the targeted range.

Applications

Some nanotechnology-based drugs that are commercially available or in human clinical trials include:

- Abraxane, approved by the U.S. Food and Drug Administration (FDA) to treat breast cancer, non-small- cell lung cancer (NSCLC) and pancreatic cancer, is the nanoparticle albumin bound paclitaxel.
- Doxil was originally approved by the FDA for the use on HIV-related Kaposi's sarcoma. It is now being used to also treat ovarian cancer and multiple myeloma. The drug is encased in liposomes, which helps to extend the life of the drug that is being distributed. Liposomes are self-assembling, spherical, closed colloidal structures that are composed of lipid bilayers that surround an aqueous spacc. The liposomes also help to increase the functionality and it helps to decrease the damage that the drug does to the heart muscles specifically.
- Onivyde, liposome encapsulated irinotecan to treat metastatic pancreatic cancer, was approved by FDA in October 2015.
- C-dots (Cornell dots) are the smallest silica-based nanoparticles with the size <10 nm. The particles are infused with organic dye which will light up with fluorescence. Clinical trial is underway since 2011 to use the C-dots as diagnostic tool to assist surgeons to identify the location of tumor cells.
- An early phase clinical trial using the platform of 'Minicell' nanoparticle for drug delivery have been tested on patients with advanced and untreatable cancer. Built from the membranes of mutant bacteria, the minicells were loaded with paclitaxel and coated with cetuximab, antibodies that bind the epidermal growth factor receptor (EGFR) which is often overexpressed in a number of

cancers, as a 'homing' device to the tumor cells. The tumor cells recognize the bacteria from which the minicells have been derived, regard it as invading microorganism and engulf it. Once inside, the payload of anti-cancer drug kills the tumor cells. Measured at 400 nanometers, the minicell is bigger than synthetic particles developed for drug delivery. The researchers indicated that this larger size gives the minicells a better profile in side-effects because the minicells will preferentially leak out of the porous blood vessels around the tumor cells and do not reach the liver, digestive system and skin. This Phase 1 clinical trial demonstrated that this treatment is well tolerated by the patients. As a platform technology, the minicell drug delivery system can be used to treat a number of different cancers with different anti-cancer drugs with the benefit of lower dose and less side-effects.

- In 2014, a Phase 3 clinical trial for treating inflammation and pain after cataract surgery, and a Phase 2 trial for treating dry eye disease were initiated using nanoparticle loteprednol etabonate. In 2015, the product, KPI-121 was found to produce statistically significant positive results for the post-surgery treatment.

Cancer

Existing and potential drug nanocarriers have been reviewed.

Nanoparticles have high surface area to volume ratio. This allows for many functional groups to be attached to a nanoparticle, which can seek out and bind to certain tumor cells. Additionally, the small size of nanoparticles (10 to 100 nanometers), allows them to preferentially accumulate at tumor sites (because tumors lack an effective lymphatic drainage system). Limitations to conventional cancer chemotherapy include drug resistance, lack of selectivity, and lack of solubility. Nanoparticles have the potential to overcome these problems.

In photodynamic therapy, a particle is placed within the body and is illuminated with light from the outside. The light gets absorbed by the particle and if the particle is metal, energy from the light will heat the particle and surrounding tissue. Light may also be used to produce high energy oxygen molecules which will chemically react with and destroy most organic molecules that are next to them (like tumors). This therapy is appealing for many reasons. It does not leave a "toxic trail" of reactive molecules throughout the body (chemotherapy) because it is directed where only the light is shined and the particles exist. Photodynamic therapy has potential for a noninvasive procedure for dealing with diseases, growth and tumors. Kanzius RF therapy is one example of such therapy (nanoparticle hyperthermia) . Also, gold nanoparticles have the potential to join numerous therapeutic functions into a single platform, by targeting specific tumor cells, tissues and organs.

Visualization

In vivo imaging is another area where tools and devices are being developed. Using nanoparticle contrast agents, images such as ultrasound and MRI have a favorable distribution and improved contrast. This might be accomplished by self assembled biocompatible nanodevices that will detect, evaluate, treat and report to the clinical doctor automatically.

The small size of nanoparticles endows them with properties that can be very useful in oncology, particularly in imaging. Quantum dots (nanoparticles with quantum confinement properties, such as size-tunable light emission), when used in conjunction with MRI (magnetic resonance imaging), can produce exceptional images of tumor sites. Nanoparticles of cadmium selenide (quantum dots) glow when exposed to ultraviolet light. When injected, they seep into cancer tumors. The surgeon can see the glowing tumor, and use it as a guide for more accurate tumor removal.These nanoparticles are much brighter than organic dyes and only need one light source for excitation. This means that the use of fluorescent quantum dots could produce a higher contrast image and at a lower cost than today's organic dyes used as contrast media. The downside, however, is that quantum dots are usually made of quite toxic elements.

Tracking movement can help determine how well drugs are being distributed or how substances are metabolized. It is difficult to track a small group of cells throughout the body, so scientists used to dye the cells. These dyes needed to be excited by light of a certain wavelength in order for them to light up. While different color dyes absorb different frequencies of light, there was a need for as many light sources as cells. A way around this problem is with luminescent tags. These tags are quantum dots attached to proteins that penetrate cell membranes. The dots can be random in size, can be made of bio-inert material, and they demonstrate the nanoscale property that color is size-dependent. As a result, sizes are selected so that the frequency of light used to make a group of quantum dots fluoresce is an even multiple of the frequency required to make another group incandesce. Then both groups can be lit with a single light source. They have also found a way to insert nanoparticles into the affected parts of the body so that those parts of the body will glow showing the tumor growth or shrinkage or also organ trouble.

Sensing

Nanotechnology-on-a-chip is one more dimension of lab-on-a-chip technology. Magnetic nanoparticles, bound to a suitable antibody, are used to label specific molecules, structures or microorganisms. Gold nanoparticles tagged with short segments of DNA can be used for detection of genetic sequence in a sample. Multicolor optical coding for biological assays has been achieved by embedding different-sized quantum dots into polymeric microbeads. Nanopore technology for analysis of nucleic acids converts strings of nucleotides directly into electronic signatures.

Sensor test chips containing thousands of nanowires, able to detect proteins and other biomarkers left behind by cancer cells, could enable the detection and diagnosis of cancer in the early stages from a few drops of a patient's blood.Nanotechnology is helping to advance the use of arthroscopes, which are pencil-sized devices that are used in surgeries with lights and cameras so surgeons can do the surgeries with smaller incisions. The smaller the incisions the faster the healing time which is better for the patients. It is also helping to find a way to make an arthroscope smaller than a strand of hair.

Research on nanoelectronics-based cancer diagnostics could lead to tests that can be done in pharmacies. The results promise to be highly accurate and the product promises to be inexpensive. They could take a very small amount of blood and detect cancer anywhere in the body in about five minutes, with a sensitivity that is a thousand times better than in a conventional laboratory test. These devices that are built with nanowires to detect cancer proteins; each nanowire detector is primed to be sensitive to a different cancer marker. The biggest advantage of the nanowire detectors is that they could test for anywhere from ten to one hundred similar medical conditions without adding cost to the testing device. Nanotechnology has also helped to personalize oncology for the detection, diagnosis, and treatment of cancer. It is now able to be tailored to each individual's tumor for better performance. They have found ways that they will be able to target a specific part of the body that is being affected by cancer.

Blood Purification

Magnetic micro particles are proven research instruments for the separation of cells and proteins from complex media. The technology is available under the name Magnetic-activated cell sorting or Dynabeads among others. More recently it was shown in animal models that magnetic nanoparticles can be used for the removal of various noxious compounds including toxins, pathogens, and proteins from whole blood in an extracorporeal circuit similar to dialysis. In contrast to dialysis, which works on the principle of the size related diffusion of solutes and ultrafiltration of fluid across a semi-permeable membrane, the purification with nanoparticles allows specific targeting of substances. Additionally larger compounds which are commonly not dialyzable can be removed.

The purification process is based on functionalized iron oxide or carbon coated metal nanoparticles with ferromagnetic or superparamagnetic properties. Binding agents such as proteins, antibodies, antibiotics, or synthetic ligands are covalently linked to the particle surface. These binding agents are able to interact with target species forming an agglomerate. Applying an external magnetic field gradient allows exerting a force on the nanoparticles. Hence the particles can be separated from the bulk fluid, thereby cleaning it from the contaminants.

The small size (< 100 nm) and large surface area of functionalized nanomagnets leads to advantageous properties compared to hemoperfusion, which is a clinically used technique for the purification of blood and is based on surface adsorption. These advantages are high loading and accessibility of the binding agents, high selectivity towards the target compound, fast diffusion, small hydrodynamic resistance, and low dosage.

This approach offers new therapeutic possibilities for the treatment of systemic infections such as sepsis by directly removing the pathogen. It can also be used to selectively remove cytokines or endotoxins or for the dialysis of compounds which are not accessible by traditional dialysis methods. However the technology is still in a preclinical phase and first clinical trials are not expected before 2017.

Tissue Engineering

Nanotechnology may be used as part of tissue engineering to help reproduce or repair or reshape damaged tissue using suitable nanomaterial-based scaffolds and growth factors. Tissue engineering if successful may replace conventional treatments like organ transplants or artificial implants. Nanoparticles such as graphene, carbon nanotubes, molybdenum disulfide and tungsten disulfide are being used as reinforcing agents to fabricate mechanically strong biodegradable polymeric nanocomposites for bone tissue engineering applications. The addition of these nanoparticles in the polymer matrix at low concentrations (~0.2 weight %) leads to significant improvements in the compressive and flexural mechanical properties of polymeric nanocomposites. Potentially, these nanocomposites may be used as a novel, mechanically strong, light weight composite as bone implants.

For example, a flesh welder was demonstrated to fuse two pieces of chicken meat into a single piece using a suspension of gold-coated nanoshells activated by an infrared laser. This could be used to weld arteries during surgery. Another example is nanonephrology, the use of nanomedicine on the kidney.

Medical Devices

Neuro-electronic interfacing is a visionary goal dealing with the construction of nanodevices that will permit computers to be joined and linked to the nervous system. This idea requires the building of a molecular structure that will permit control and detection of nerve impulses by an external computer. A refuelable strategy implies energy is refilled continuously or periodically with external sonic, chemical, tethered, magnetic, or biological electrical sources, while a nonrefuelable strategy implies that all power is drawn from internal energy storage which would stop when all energy is drained. A nanoscale enzymatic biofuel cell for self-powered nanodevices have been developed that uses glucose from biofluids including human blood and watermelons. One limitation to this innovation is the fact that electrical interference or leakage or overheating from power consumption is possible. The wiring of the structure is extremely difficult because they must be positioned precisely in the nervous system. The structures that will provide the interface must also be compatible with the body's immune system.

Molecular nanotechnology is a speculative subfield of nanotechnology regarding the possibility of engineering molecular assemblers, machines which could re-order matter at a molecular or atomic scale. Nanomedicine would make use of these nanorobots, introduced into the body, to repair or detect damages and infections. Molecular nanotechnology is highly theoretical, seeking to anticipate what inventions nanotechnology might yield and to propose an agenda for future inquiry. The proposed elements of molecular nanotechnology, such as molecular assemblers and nanorobots are far beyond current capabilities. Future advances in nanomedicine could give rise to life extension through the repair of many processes thought to be responsible for aging. K. Eric Drexler, one of the founders of nanotechnology, postulated cell repair machines, including ones operating within cells and utilizing as yet hypothetical molecular machines, in his 1986 book Engines of Creation, with the first technical discussion of medical nanorobots by Robert Freitas appearing in 1999. Raymond Kurzweil, a futurist and transhumanist, stated in his book *The Singularity Is Near* that he believes that advanced medical nanorobotics could completely remedy the effects of aging by 2030. According to Richard Feynman, it was his former graduate student and collaborator Albert Hibbs who originally suggested to him (circa 1959) the idea of a *medical* use for Feynman's theoretical micromachines (see nanotechnology). Hibbs suggested that certain repair machines might one day be reduced in size to the point that it would, in theory, be possible to (as Feynman put it) "swallow the doctor". The idea was incorporated into Feynman's 1959 essay *There's Plenty of Room at the Bottom.*

REFERENCES

1. "Nanosystems Glossary". E-drexler.com.
2. "Doing MM". Wise Nano. 2008-09-24. Retrieved 2010-09-05.
3. "Foresight Institute press release". Foresight.org. 2008-01-29. Retrieved 2010-09-05.
4. Peterson, Christine (2007-05-08). "Nanodot: Nanotechnology News and Discussion " Blog Archive " Nanotechnology Roadmap launch: Productive Nanosystems Conference, Oct 9-10". Foresight.org. Retrieved 2010-09-05.
5. "Nanofactory Collaboration". Molecularassembler.com. Retrieved 2010-09-05.
6. "Nanofactory Technical Challenges". Molecularassembler.com. Retrieved 2010-09-05.
7. "Global Task Force on Implications and Policy". Crnano.org. Retrieved 2010-09-05.
8. "3.23.4". Molecularassembler.com. 2005-08-01. Retrieved 2010-09-05.
9. Molecular Nanotechnology Guidelines". Foresight.org. Retrieved 2010-09-05.
10. "5.1.9". Molecularassembler.com. 2005-08-01. Retrieved 2010-09-05.
11. "N04FR06-p.15.pmd" (PDF). Retrieved 2010-09-05.
12. "NanomedicineBookSite". Nanomedicine.com. Retrieved 2010-09-05.
13. "NanoPubls". Rfreitas.com. Retrieved 2010-09-05.
14. http://discuss.foresight.org/~josh/Ufog.html
15. "Phased Array Optics". Phased-array.com. Retrieved 2010-09-05.
16. "Phased Array Optics". Phased-array.com. 1991-10-03. Retrieved 2010-09-05.
17. "Frequently Asked Questions - Molecular Manufacturing". foresight.org. Retrieved 19 July 2014.
18. Chris Phoenix; Mike Treder (2008). "Chapter 21: Nanotechnology as global catastrophic risk". In Bostrom, Nick; Cirkovic, Milan M. Global catastrophic risks. Oxford: Oxford University Press. ISBN 978-0-19-857050-9.
19. Sandberg, Anders. "The five biggest threats to human existence". http://theconversation.com/. Retrieved 13 July 2014. External link in |website= (help)
20. Drexler, Eric. "A Dialog on Dangers". foresight.org. Retrieved 19 July 2014.
21. Drexler, Eric. "ENGINES OF DESTRUCTION (Chapter 11)". http://e-drexler.com/. Retrieved 19 July 2014.External link in |website= (help)
22. Tomasik, Brian. "Possible Ways to Promote Compromise". http://foundational-research.org/. Retrieved 19 July 2014. External link in |website= (help)
23. «Dangers of Molecular Manufacturing". crnano.org. Retrieved 19 July 2014.

24. The Need for International Control". crnano.org. Retrieved 19 July 2014.
25. Tomasik, Brian. "International Cooperation vs. AI Arms Race". foundational-research.org. Retrieved 19 July 2014.
26. "Technical Restrictions May Make Nanotechnology Safer". crnano.org. Retrieved 19 July 2014.
27. Tomasik, Brian. "Possible Ways to Promote Compromise". http://foundational-research.org/. Retrieved 22 July 2014. External link in |website= (help)
28. Joseph, Lawrence E. (2007). Apocalypse 2012. New York: Broadway. p. 6. ISBN 978-0-7679-2448-1.
29. Rincon, Paul (2004-06-09). "Nanotech guru turns back on ‹goo'". BBC News. Retrieved 2012-03-30.
30. Hapgood, Fred (November 1986). "Nanotechnology: Molecular Machines that Mimic Life" (PDF). Omni. Retrieved 19 July 2014.
31. "Leading nanotech experts put ‹grey goo' in perspective". crnano.org. Retrieved 19 July 2014.
32. Merkle, Ralph (22 June 1999). "Nanotechnology: the coming revolution in manufacturing, Testimony to the U.S. House of Representatives Committee on Science, Subcommittee on Basic Research".
33. "N20FR06-p._.pmd" (PDF). Retrieved 2010-09-05.
34. http://lifeboat.com/ex/corporate.cornucopia
35. "Tangible Nanomoney". Rfreitas.com. Retrieved 2010-09-05.
36. "Nanotechnology: Dangers of Molecular Manufacturing". Crnano.org. Retrieved 2010-09-05.
37. "Some Limits to Global Ecophagy". Rfreitas.com. Retrieved 2010-09-05.
38. "Foresight Guidelines on Molecular Nanotechnology". Foresight.org. 2006-04-06. Retrieved 2010-09-05.
39. A Matter of Size: Triennial Review of the National Nanotechnology Initiative". Nap.edu. Retrieved 2010-09-05.
40. "Engines of Creation - K. Eric Drexler : Cover". E-drexler.com. Retrieved 2010-09-05.
41. "How good scientists reach bad conclusions". Foresight.org. Retrieved 2010-09-05.
42. "Nanosystems TOC". E-drexler.com. 2002-11-01. Retrieved 2010-09-05.
43. Smalley, Richard E. (September 2001). "Of Chemistry, Love and Nanobots". Scientific American.
44. "Debate About Assemblers — Smalley Rebuttal". Imm.org. Retrieved 2010-09-05.
45. "C&En: Cover Story - Nanotechnology". Pubs.acs.org. 2003-12-01. Retrieved 2010-09-05.
46. Gall, John, (1986) Systemantics: How Systems Really Work and How They

Fail, 2nd ed. Ann Arbor, MI : The General Systemantics Press.

47. Richard Dawkins, The Blind Watchmaker: Why the Evidence of Evolution Reveals a Universe Without Design, W. W. Norton; Reissue edition (September 19, 1996)

48. "Blog Archive " Is mechanosynthesis feasible? The debate moves up a gear". Soft Machines. 2004-12-16. Retrieved 2010-09-05.

49. "Smalley". Wired.com. Retrieved 2010-09-05.

50. "Atkinson". Nanotech-now.com. Retrieved 2010-09-05.

51. "Moriarty". Softmachines.org. 2005-01-26. Retrieved 2010-09-05.

52. "Jones". Softmachines.org. 2005-12-18. Retrieved 2010-09-05.

53. "Nanofactory Collaboration Publications". Molecularassembler.com. Retrieved 2010-09-05.

54. "Moriarity Correspondence" (PDF). Retrieved 2010-09-05.

55. "Nanofactory Collaboration". Molecularassembler.com. Retrieved 2010-09-05.

56. "Nanorex, Inc. - Molecular Machinery Gallery". Nanoengineer-1.com. Retrieved 2010-09-05.

57. "Diamond Mechanosynthesis". Molecularassembler.com. Retrieved 2010-09-05.

58. "Technology Roadmap for Productive Nanosystems". Foresight.org. Retrieved 2010-09-05.

59. Wikiquote:Richard Feynman

60. High-level Ab Initio Studies of Hydrogen Abstraction from Prototype Hydrocarbon Systems. Temelso, Sherrill, Merkle, and Freitas, J. Phys. Chem. A Vol. 110, pages 11160-11173, 2006.

61. Theoretical Analysis of a Carbon-Carbon Dimer Placement Tool for Diamond Mechanosynthesis. Merkle and Freitas, J. Nanosci. Nanotech. Vol. 3, pages 319-324, 2003.

62. Theoretical Analysis of Diamond Mechanosynthesis. Part I. Stability of C_2 Mediated Growth of Nanocrystalline Diamond C(110) Surface. Peng, Freitas and Merkle. J. Comput. Theor. Nanosci. Vol. 1, pages 62-70, 2004.

63. Theoretical Analysis of Diamond Mechanosynthesis. Part II. C_2 Mediated Growth of Diamond C(110) Surface via Si/Ge-Triadamantane Dimer Placement Tools. Mann, Peng, Freitas and Merkle. J. Comput. Theor. Nanosci. Vol. 1, pages 71-80, 2004.

64. Design and Analysis of a Molecular Tool for Carbon Transfer in Mechanosynthesis. Allis and Drexler. J. Comput. Theor. Nanosci. Vol. 2, pages 71-80, 2005.

65. Theoretical Analysis of Diamond Mechanosynthesis. Part III. Positional C_2 Deposition on Diamond C(110) Surface using Si/Ge/Sn-based Dimer Placement Tools. Peng, Freitas, Merkle, Von Ehr, Randall and Skidmore. J. Comput. Theor. Nanosci. Vol. 3, pages 28-41, 2006.

66. [Horizontal Ge-Substituted Polymantane-Based C_2 Dimer Placement Tooltip

Motifs for Diamond Mechanosynthesis]. Freitas, Allis and Merkle. J. Comput. Theor. Nanosci. Vol. 4, 2007, in press.

67. "03CTN01-003" (PDF). Retrieved 2010-09-05.

68. "Wilson Ho". Physics.uci.edu. Retrieved 2010-09-05.

69. "Mechanical Vertical Manipulation of Selected Single Atoms by Soft Nanoindentation Using Near Contact Atomic Force Microscopy". Physical Review Letters. 90: 176102. Bibcode:2003PhRvL..90q6102O. doi:10.1103/PhysRevLett.90.176102.

70. R. V. Lapshin (2004). "Feature-oriented scanning methodology for probe microscopy and nanotechnology"(PDF). Nanotechnology. UK: IOP. 15 *(9): 1135–1151. doi:10.1088/0957-4484/15/9/006. ISSN 0957-4484.*

71. R. V. Lapshin (2011). "Feature-oriented scanning probe microscopy". In H. S. Nalwa. Encyclopedia of Nanoscience and Nanotechnology (PDF). 14. USA: American Scientific Publishers. pp. 105–115. ISBN 1-58883-163-9.

72. «DMS Bibliography". Molecularassembler.com. Retrieved 2010-09-05.

CHAPTER 9

Nanomechanics

Chapter Contents

Nanomechanics is the branch of nanotechnology that directly deals with the mechanical, functional and elastic characteristics of the different kind of systems having nanometer material involved in them. Its one of the field that serves as the foundation of nanotechnology and nanoscience. It also handles the thermal and kinetic properties of the materials.

INTRODUCTION

Nanomechanics is a branch of *nanoscience* studying fundamental *mechanical* (elastic, thermal and kinetic) properties of physical systems at the nanometer scale. Nanomechanics has emerged on the crossroads of classical mechanics, solid-state physics, statistical mechanics, materials science, and quantum chemistry. As an area of nanoscience, nanomechanics provides a scientific foundation of nanotechnology.

Nanomechanics is that branch of nanoscience which deals with the study and application of fundamental mechanical properties of physical systems at the nanoscale, such as elastic, thermal and kinetic material properties.

Often, nanomechanics is viewed as a *branch* of nanotechnology, i.e., an applied area with a focus on the mechanical properties of *engineered* nanostructures and nanosystems (systems with nanoscale components of importance). Examples of the latter include nanoparticles, nanopowders, nanowires, nanorods, nanoribbons, nanotubes, including carbon nanotubes (CNT) and boron nitride nanotubes (BNNTs); nanoshells, nanomebranes, nanocoatings, nanocomposite/nanostructured materials, (fluids with dispersed nanoparticles); nanomotors, etc. Some of the well-established *fields of nanomechanics* are: nanomaterials, nanotribology (friction, wear and contact mechanics at the nanoscale), nanoelectromechanical systems (NEMS), and nanofluidics. As a fundamental science, nanomechanics is based on some empirical principles (basic observations), namely general mechanics principles and specific principles arising from the smallness of physical sizes of the object of study.

General mechanics principles include:

- Energy and momentum conservation principles
- Variational Hamilton's principle
- Symmetry principles

Due to smallness of the studied object, nanomechanics also accounts for:

- *Discreteness* of the object, whose size is comparable with the interatomic distances
- Plurality, but *finiteness*, of degrees of freedom in the object
- Importance of *thermal fluctuations*
- Importance of *entropic effects* (see configuration entropy)
- Importance of *quantum effects* (see quantum machine)

These principles serve to provide a basic insight into novel mechanical properties of nanometer objects. Novelty is understood in the sense that these properties are not present in similar macroscale objects or much different from the properties of those (e.g., nanorods vs. usual macroscopic beam structures). In particular, smallness of the subject itself gives rise to various surface effects determined by higher surface-to-volume ratio of nanostructures, and thus affects mechanoenergetic and thermal properties (melting point, heat capacitance, etc.) of nanostructures. Discreteness serves a fundamental reason, for instance, for the dispersion of mechanical waves in solids, and some special behavior of basic elastomechanics solutions at small scales. Plurality of degrees of freedom and the rise of thermal fluctuations are the reasons for thermal tunneling of nanoparticles through potential barriers, as well as for the cross-diffusion of liquids and solids. Smallness and thermal fluctuations provide the basic reasons of the Brownian motion of nanoparticles. Increased importance of thermal fluctuations and configuration entropy at the nanoscale give rise to superelasticity, entropic elasticity (entropic forces), and other exotic types of elasticity of nanostructures. Aspects of configuration entropy are also of great interest in the context self-organization and cooperative behavior of open nanosystems.

Quantum effects determine *forces of interaction* between individual atoms in physical objects, which are introduced in nanomechanics by means of some averaged mathematical models called *interatomic potentials*.

Subsequent utilization of the interatomic potentials within the classical multibody dynamics provide deterministic mechanical models of nano structures and systems at the atomic scale/resolution. Numerical methods of solution of these models are called *molecular dynamics* (MD), and sometimes *molecular mechanics* (especially, in relation to statically equilibrated (still) models). Non-deterministic numerical approaches include

Monte Carlo, Kinetic More-Carlo (KMC), and other methods. Contemporary numerical tools include also hybrid *multiscale approaches* allowing concurrent or sequential utilization of the atomistic scale methods (usually, MD) with the continuum (macro) scale methods (usually, field emission microscopy) within a single mathematical model. Development of these complex methods is a separate subject of applied mechanics research.

Quantum effects also determine novel electrical, optical and chemical properties of nanostructures, and therefore they find even greater attention in adjacent areas of nanoscience and nanotechnology, such as nanoelectronics, advanced energy systems, and nanobiotechnology.

MOLECULAR MACHINE

A **molecular machine**, or **nanomachine**, refers to any discrete number of molecular components that produce quasi-mechanical movements (output) in response to specific stimuli (input). The expression is often more generally applied to molecules that simply mimic functions that occur at the macroscopic level. The term is also common in nanotechnology where a number of highly complex molecular machines have been proposed that are aimed at the goal of constructing a molecular assembler. Molecular machines can be divided into two broad categories; synthetic and biological.

The 2016 Nobel Prize in Chemistry was awarded to Jean-Pierre Sauvage, Sir J. Fraser Stoddart and Bernard L. Feringa for the design and synthesis of molecular machines.

Examples

From a synthetic perspective, there are two important types of molecular machines: molecular switches (or shuttles) and molecular motors. The major difference between the two systems is that a switch influences a system as a function of state, whereas a motor influences a system as a function of trajectory. A switch (or shuttle) may appear to undergo translational motion, but returning a switch to its original position undoes any mechanical effect and liberates energy to the system. Furthermore, switches cannot use chemical energy to repetitively and progressively drive a system away from equilibrium whereas a motor can.

Synthetic

A wide variety of rather simple molecular machines have been synthesized by chemists. They can consist of a single molecule; however, they are often constructed for mechanically-interlocked molecular architectures, such as rotaxanes and catenanes. Carbon nanotube nanomotors have also been produced.

- Molecular motors are molecules that are capable of unidirectional rotation motion powered by external energy input. A number of molecular machines have been synthesized powered by light or reaction with other molecules.
- A molecular propeller is a molecule that can propel fluids when rotated, due to its special shape that is designed in analogy to macroscopic propellers. It

has several molecular-scale blades attached at a certain pitch angle around the circumference of a nanoscale shaft. Also see molecular gyroscope.

- A molecular switch is a molecule that can be reversibly shifted between two or more stable states. The molecules may be shifted between the states in response to changes in e.g. pH, light, temperature, an electric current, microenvironment, or the presence of a ligand.
- A molecular shuttle is a molecule capable of shuttling molecules or ions from one location to another. A common molecular shuttle consists of a rotaxane where the macrocycle can move between two sites or stations along the dumbbell backbone.
- Molecular tweezers are host molecules capable of holding items between their two arms. The open cavity of the molecular tweezers binds items using non-covalent bonding including hydrogen bonding, metal coordination, hydrophobic forces, van der Waals forces, π interactions, or electrostatic effects. Examples of molecular tweezers have been reported that are constructed from DNA and are considered DNA machines.
- A molecular sensor is a molecule that interacts with an analyte to produce a detectable change. Molecular sensors combine molecular recognition with some form of reporter, so the presence of the item can be observed.
- A molecular logic gate is a molecule that performs a logical operation on one or more logic inputs and produces a single logic output. Unlike a molecular sensor, the molecular logic gate will only output when a particular combination of inputs are present.

Biological

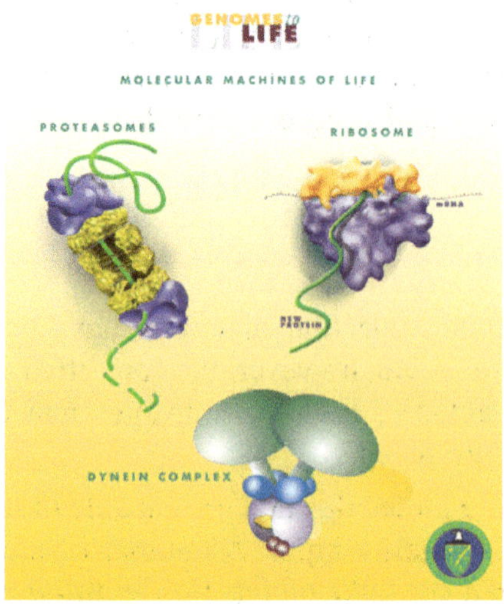

Some biological molecular machines./

The most complex molecular machines are proteins found within cells. These include motor proteins, such as myosin, which is responsible for muscle contraction, kinesin, which moves cargo inside cells away from the nucleus along microtubules, and dynein, which produces the axonemal beating of motile cilia and flagella. These proteins and their nanoscale dynamics are far more complex than any molecular machines that have yet been artificially constructed. Probably the most significant biological machine known is the ribosome. Other important examples include motile cilia. A high-level-abstraction summary is that, "[i]n effect, the [motile cilium] is a nanomachine composed of perhaps over 600 proteins in molecular complexes, many of which also function independently as nanomachines." Flexible linker domains allow the connecting protein domains to recruit their binding partners and induce long-range allostery via protein domain dynamics.

This protein flexibility allows the construction of biological machines. The first useful applications of these biological machines might be in nanomedicine. For example, they could be used to identify and destroy cancer cells. Molecular nanotechnology is a speculative subfield of nanotechnology regarding the possibility of engineering molecular assemblers, biological machines which could re-order matter at a molecular or atomic scale. Nanomedicine would make use of these nanorobots, introduced into the body, to repair or detect damages and infections. Molecular nanotechnology is highly theoretical, seeking to anticipate what inventions nanotechnology might yield and to propose an agenda for future inquiry. The proposed elements of molecular nanotechnology, such as molecular assemblers and nanorobots are far beyond current capabilities.

Research

The construction of more complex molecular machines is an active area of theoretical research. A number of molecules, such as molecular propellers, have been designed, although experimental studies of these molecules are inhibited by the lack of methods to construct these molecules. In this context, theoretical modeling can be extremely useful to understand the self-assembly/disassembly processes of rotaxanes, important for the construction of light-powered molecular machines. This molecular-level knowledge may foster the realization of ever more complex, versatile, and effective molecular machines for the areas of nanotechnology, including molecular assemblers.

NANOELECTROMECHANICAL RELAY

A **nanoelectromechanical (NEM) Relay** is an electrically actuated switch that is built on the nanometer scale using semiconductor fabrication techniques. They are designed to operate in replacement, or in conjunction, with traditional semiconductor logic. While the mechanical nature of NEM relays makes them switch much slower than solid-state relays, they have many advantageous properties, such as zero current leakage and low power consumption, which make them potentially useful in next generation computing.

A typical NEM Relay requires a potential on the order of the tens of volts in order to "pull in" and have contact resistances on the order of gigaohms. Coating contact surfaces with platinum can reduce achievable contact resistance to as low as 3 kΩ. Compared to transistors, NEM relays switch relatively slowly, on the order of nanoseconds.

Operation

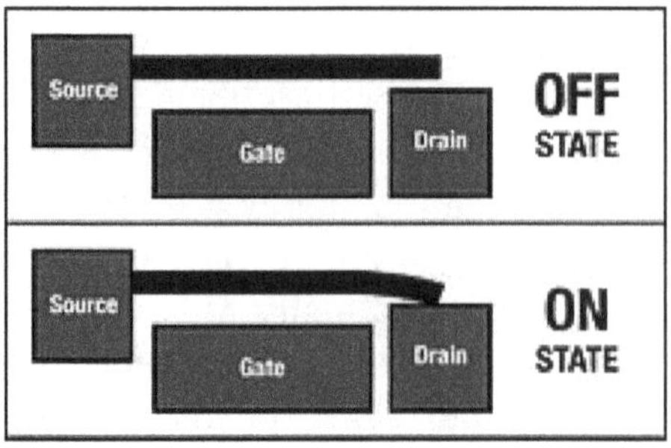

Schematic of a three terminal electromechanical relay

A NEM relay can be fabricated in two, three, or four terminal configurations. A three terminal relay is composed of a source (input), drain (output), and a gate (actuation terminal). Attached to the source is a cantilevered beam that can be bent into contact with the drain in order to make an electrical connection. When a significant voltage differential is applied between the beam and gate, and the electrostatic force overcomes the elastic force of the beam enough to bend it into contact with the drain, the device "pulls in" and forms an electrical connection. In the off position, the source and drain are separated by an air gap. This physical separation allows NEM relays to have zero current leakage, and very sharp on/off transitions.

The nonlinear nature of the electric field, and adhesion between the beam and drain cause the device to "pull out" and lose connection at a lower voltage than the voltage at which it pulls in. This hysteresis effect means there is a voltage between the pull in voltage, and the pull out voltage that will not change the state of the relay, no matter what its initial state is. This property is very useful in applications where information needs to be stored in the circuit, such as in static random-access memory.

Fabrication

NEM relays are usually fabricated using surface micromachining techniques typical of microelectromechanical systems (MEMS). Laterally actuated relays are constructed by first depositing two or more layers of material on a silicon wafer. The upper structural layer is photolithographically patterned in order to form isolated blocks of the uppermost material. The layer below is then selectively etched away, leaving thin structures, such as the relay's beam, cantilevered above the wafer, and free to bend laterally. A common set of materials used in this process is polysilicon as the upper structural layer, and silicon dioxide as the sacrificial lower layer.

NEM relays can be fabricated using a back end of line compatible process, allowing them to be built on top of CMOS. This property allows NEM relays to be used to significantly reduce the area of certain circuits. For example, a CMOS-NEM relay hybrid inverter occupies 0.03 μm^2, one-third the area of a 45 nm CMOS inverter.

The first switch made using silicon micro-machining techniques was fabricated in 1978. Those switches were made using bulk micromachining processes and electroplating. In the 1980s, surface micromachining techniques were developed and the technology was applied to the fabrication of switches, allowing for smaller, more efficient relays.

A major early application of MEMS relays was for switching radio frequency signals at which solid-state relays had poor performance. The switching time for these early relays was above 1 µs. By shrinking dimensions below one micrometer, and moving into the nano scale, MEMS switches have achieved switching times in the ranges of hundreds of nanoseconds.

Applications

Mechanical Computing

Due to transistor leakage, there is a limit to the theoretical efficiency of CMOS logic. This efficiency barrier ultimately prevents continued increases in computing power in power-constrained applications. While NEM relays have significant switching delays, their small size and fast switching speed when compared to other relays means that mechanical computing utilizing NEM Relays could prove a viable replacement for typical CMOS based integrated circuits, and break this CMOS efficiency barrier.

A NEM relay switches mechanically about 1000 times slower than a solid-state transistor takes to switch electrically. While this makes using NEM relays for computing a significant challenge, their low resistance would allow many NEM relays to be chained together and switch all at once, performing a single large calculation. On the other hand, transistor logic has to be implemented in small cycles of calculations, because their high resistance does not allow many transistors to be chained together while maintaining signal integrity. Therefore, it would be possible to create a mechanical computer using NEM relays that operates at a much lower clock speed than CMOS logic, but performs larger, more complex calculations during each cycle. This would allow a NEM relay based logic to perform to standards comparable to current CMOS logic.

There are many applications, such as in the automotive, aerospace, or geothermal exploration businesses, in which it would be beneficial to have a microcontroller that could operate at very high temperatures. However, at high temperatures, semiconductors used in typical microcontrollers begin to fail as the electrical properties of the materials they are made of degrade, and the transistors no longer function. NEM relays do not rely on the electrical properties of materials to actuate, so a mechanical computer utilizing NEM relays would be able to operate in such conditions. NEM relays have been successfully tested at up to 500 °C, but could theoretically withstand much higher temperatures.

Field-Programmable Gate Arrays

The zero leakage current, low energy usage, and ability to be layered on top of CMOS properties of NEM relays make them a promising candidate for usage as routing switches in Field-programmable gate arrays (FPGA). A FPGA utilizing a NEM relay to replace each routing switch and its corresponding static random-access memory block could allow for a significant reduction in programming delay, power leakage, and chip area compared to a typical 22nm CMOS based FPGA. This area reduction mainly comes from the fact that the NEM relay routing layer can be built on top of the CMOS layer of the FPGA.

SOLID-STATE RELAY

Solid-state relay. ~240v 2.5A.

A solid-state relay (SSR) is an electronic switching device that switches on or off when a small external voltage is applied across its control terminals. SSRs consist of a sensor which responds to an appropriate input (control signal), a solid-state electronic switching device which switches power to the load circuitry, and a coupling mechanism to enable the control signal to activate this switch without mechanical parts. The relay may be designed to switch either AC or DC to the load. It serves the same function as an electromechanical relay, but has no moving parts.

Packaged solid-state relays use power semiconductor devices such as thyristors and transistors, to switch currents up to around a hundred amperes. Solid-state relays have fast switching speeds compared with electromechanical relays, and have no physical contacts to wear out. Application of solid-state relays must consider their lower ability to withstand momentary overload, compared with electromechanical contacts, and their higher "on" state resistance. Unlike an electromechanical relay, a solid-state relay provides only limited switching arrangements (SPST switching).

Coupling

The control signal must be coupled to the controlled circuit in a way which provides galvanic isolation between the two circuits.

Many SSRs use optical coupling. The control voltage energizes an internal LED which illuminates and switches on a photo-sensitive diode (photo-voltaic); the diode current turns on a back-to-back thyristor, SCR, or MOSFET to switch the load. The optical coupling allows the control circuit to be electrically isolated from the load.

Operation

An SSR based on a single MOSFET, or multiple MOSFETs in a paralleled array, can work well for DC loads. MOSFETs have an inherent substrate diode that conducts in the reverse direction, so a single MOSFET cannot block current in both directions. For AC (bi-directional) operation two MOSFETs are arranged back-to-back with their source pins tied together. Their drain pins are connected to either side of the output. The substrate diodes are alternately reverse biased to block current when the relay is off. When the relay is on, the common source is always riding on the instantaneous signal level and both gates are biased positive relative to the source by the photo-diode.

It is common to provide access to the common source so that multiple MOSFETs can be wired in parallel if switching a DC load. Usually a network is provided to speed the turn-off of the MOSFET when the control input is removed.

In AC circuits, SCR or TRIAC relays inherently switch off at the points of zero load current. The circuit will never be interrupted in the middle of a sine wave peak, preventing the large transient voltages that would otherwise occur due to the sudden collapse of the magnetic field around the inductance. This feature is called zero-crossover switching.

Parameters

SSRs are characterised by a number of parameters including the required activating input voltage, current, output voltage and current, whether it is AC or DC, voltage drop or resistance affecting output current, thermal resistance, and thermal and electrical parameters for safe operating area (e.g., derating according to thermal resistance when repeatedly switching large currents).

Advantages over Mechanical Relays

Most of the relative advantages of solid state and electromechanical relays are common to all solid-state as against electromechanical devices.

- Slimmer profile, allowing tighter packing.
- Totally silent operation.
- SSRs switch faster than electromechanical relays; the switching time of a typical optically coupled SSR is dependent on the time needed to power the LED on and off - of the order of microseconds to milliseconds.
- Increased lifetime, even if it is activated many times, as there are no moving parts to wear and no contacts to pit or build up carbon.
- Output resistance remains constant regardless of amount of use.

- Clean, bounceless operation.
- No sparking, allows it to be used in explosive environments, where it is critical that no spark is generated during switching.
- Inherently smaller than a mechanical relay of similar specification (if desired may have the same "casing" form factor for interchangeability).
- Much less sensitive to storage and operating environment factors such as mechanical shock, vibration, humidity, and external magnetic fields.

Disadvantages

- Voltage/current characteristic of semiconductor rather than mechanical contacts:
  - When closed, higher resistance (generating heat), and increased electrical noise
  - When open, lower resistance, and reverse leakage current (typically μA range)
  - Voltage/current characteristic is not linear (not purely resistive), distorting switched waveforms to some extent. An electromechanical relay has the low ohmic (linear) resistance of the associated mechanical switch when activated, and the exceedingly high resistance of the air gap and insulating materials when open.
  - Some types have polarity-sensitive output circuits. Electromechanical relays are not affected by polarity.
- Possibility of spurious switching due to voltage transients (due to much faster switching than mechanical relay)
- Isolated bias supply required for gate charge circuit
- Higher transient reverse recovery time (Trr) due to the presence of the body diode
- Tendency to fail "shorted" on their outputs, while electromechanical relay contacts tend to fail "open".

REFERENCES

1. Satir, Peter; Søren T. Christensen (2008-03-26). "Structure and function of mammalian cilia". Histochemistry and Cell Biology. Springer Berlin / Heidelberg. **129** *(6): 688.* doi:10.1007/s00418-008-0416-9. PMC 2386530. PMID 18365235. 1432-119X. Retrieved 2009-09-11.
2. Ballardini R, Balzani V, Credi A, Gandolfi MT, Venturi M (2001). "Artificial Molecular-Level Machines: Which Energy To Make Them Work?". *Acc. Chem. Res.* **34** *(6): 445–455.* doi:10.1021/ar000170g.
3. Staff (5 October 2016). "The Nobel Prize in Chemistry 2016". Nobel Foundation. Retrieved 5 October 2016.
4. Chang, Kenneth; Chan, Sewell (5 October 2016). "3 Makers of ‹World's Smallest Machines' Awarded Nobel Prize in Chemistry". New York Times. Retrieved 5 October

2016.

5. Fennimore, A. M.; T.D. Yuzvinsky; Wei-Qiang Han; M. S. Fuhrer; J. Cumings & A. Zettl (2003). "Rotational actuators based on carbon nanotubes". Nature. **424** *(6947): 408–410. Bibcode*:2003Natur.424..408F. doi:10.1038/nature01823. PMID 12879064.

6. Cavalcanti A, Shirinzadeh B, Freitas Jr RA, Hogg T (2008). "Nanorobot architecture for medical target identification". Nanotechnology. **19** *(1): 015103(15pp). Bibcode*:2008Nanot..19a5103C. doi:10.1088/0957-4484/19/01/015103.

7. Bu Z, Callaway DJ (2011). "Proteins MOVE! Protein dynamics and long-range allostery in cell signaling". Advances in Protein Chemistry and Structural Biology. Advances in Protein Chemistry and Structural Biology. **83**: 163–221. doi:10.1016/B978-0-12-381262-9.00005-7. ISBN 9780123812629. PMID 21570668.

8. Amrute-Nayak, M.; Diensthuber, R. P.; Steffen, W.; Kathmann, D.; Hartmann, F. K.; Fedorov, R.; Urbanke, C.; Manstein, D. J.; Brenner, B.; Tsiavaliaris, G. (2010). "Targeted Optimization of a Protein Nanomachine for Operation in Biohybrid Devices". Angewandte Chemie. **122** *(2): 322–326. doi:10.1002/ange.200905200.*

9. Patel, G. M.; Patel, G. C.; Patel, R. B.; Patel, J. K.; Patel, M. (2006). "Nanorobot: A versatile tool in nanomedicine". Journal of Drug Targeting. **14** *(2): 63–7.* doi:10.1080/10611860600612862. PMID 16608733.

10. Balasubramanian, S.; Kagan, D.; Jack Hu, C. M.; Campuzano, S.; Lobo-Castañon, M. J.; Lim, N.; Kang, D. Y.; Zimmerman, M.; Zhang, L.; Wang, J. (2011). "Micromachine-Enabled Capture and Isolation of Cancer Cells in Complex Media". Angewandte Chemie International Edition. **50** *(18): 4161–4164. doi:10.1002/anie.201100115.*

11. Freitas, Robert A., Jr.; Havukkala, Ilkka (2005). "Current Status of Nanomedicine and Medical Nanorobotics" (PDF). Journal of Computational and Theoretical Nanoscience. **2** *(4): 1–25. doi:10.1166/jctn.2005.001.*

12. Nanofactory Collaboration

13. Tabacchi, G.; Silvi, S.; Venturi, M.; Credi, A.; Fois, E. (2016). "Dethreading of a Photoactive Azobenzene-Containing Molecular Axle from a Crown Ether Ring: A Computational Investigation". ChemPhysChem. doi:10.1002/cphc.201501160.

Index